HIGH FREQUENCY ETHERNET CABLING
Analysis and Optimization

HIGH FREQUENCY ETHERNET CABLING

Analysis and Optimization

Mohamed Kwasi

KNOWLEDGE
BAKERS

KNOWLEDGE BAKERS

Cataloging in Publication Data--DK

Courtesy: D.K. Agencies (P) Ltd. <docinfo@dkagencies.com>

Kwasi, Mohamed, author.

High frequency ethernet cabling : analysis and optimization / by Mohamed Kwasi.

pages cm

Includes bibliographical references.

ISBN 9789391502096

1. Ethernet (Local area network system) I. Title.

LCC TK5105.8.E83K83 2023 | DDC 004.68 23

ISBN: 978-93-91502-09-6

Published by : **Knowledge Bakers**
Pune, India
Email: knowledgebakers2019@gmail.com

Digitally Printed at : **Replika Press Pvt. Ltd.**

Acknowledgements

I will like to thank my first supervisor, Prof. Alistair Duffy, for his advice, guidance, support, comments and inputs shared during my meetings with him that made this PhD research and thesis to be completed. My thanks also go to my second supervisor, Dr. John Gow for his support and suggestions during the period of study. Special appreciation goes to my PhD study advisor, Dr. Charles Nche, for his valuable advice, support and contributions towards the success of this thesis.

My appreciation also goes to Dr. Hugh Sasse, Ms. Virpal Kang and Ms. Florence Akinnuoye of the Electronic and Communications Laboratory, De Montfort University, Leicester, United Kingdom for their technical assistance. I will also acknowledge the efforts of Paul Cave from Mayflex and Fluke Instruments for their support in the measurements.

Finally, I will like to appreciate my wife for her care, support and encouragement throughout the period of study.

Mohamed Kwasi

CONTENTS

List of Abbreviations

40GBASE-T	40 Gigabit Ethernet
ADM	Amplitude Difference Measure
ANSI	American National Standards Institute
CCA	Copper Clad Aluminium
FDM	Feature Difference Measure
FEXT	Far-end Crosstalk
FSV	Feature Selective Validation
GA	Genetic Algorithm
GDM	Global Difference Measure
IEC	International Electrotechnical Commission
ISO	International Organization for Standardization
KS	Kolmogorov-Smirnov
LAN	Local Area Network
MAPE	Mean Absolute Prediction Error
MATLAB	Matrix Laboratory
NEXT	Near-end Crosstalk
S-parameters	Scattering parameters
TIA	Telecommunication Industry Association
A-parameters	Transfer parameters
UTP	Unshielded Twisted Pair

LIST OF FIGURES

Page

LIST OF TABLES

Chapter 1

INTRODUCTION

This chapter offers an overview of the research background and statement of problem, aims and objectives, contributions to knowledge and thesis organization.

1.1 Background and Statement of Problem

Ethernet is a family of networking technologies that was originally developed for wired Local Area Network (LAN) [1], [2]. Ethernet is now the most widely used networking technology, due largely to factors such as low cost, reliability, scalability and the availability of management tools [3], [4].

The first Ethernet standard developed in 1985 described a system that operated at 10Mb/s and specifies a half-duplex carrier with a Medium Access Control over coaxial cable [5]. This was followed by an amendment that provided specifications for fiber optic and twisted pair cabling [5]. The standardization of Ethernet technology for twisted pair cabling led to a tremendous expansion in the use of Ethernet due to the possibility of structured cabling in buildings that is of low-cost and high reliability [3].

The successes recorded in the use of twisted pair Ethernet brought about an increase in the number of users and demand for bandwidth. The increasing demand for speed led to specifications developed for fast Ethernet at 100Mb/s in 1995. Since then, successive specifications for increase in speed have been provided to 10 Gb/s (10GBASE-T) in 2006 [5].

The need for higher bandwidth to accommodate increasing demand for multimedia services led to IEEE P802.3bq task force formation [6]. The IEEE P802.3bq task force with standards bodies like the International Organization for Standardization (ISO) and the International Electrotechnical Commission (IEC) are creating standards and specifications for 40 Gb/s (40GBASE-T) Category 8 cable. This is the next generation copper twisted pair cabling over Ethernet [6]. The 40GBASE-T Ethernet is expected to offer bandwidth up to 2000MHz over twisted pair at a maximum channel length of 30m with two connectors [7].

The research in this thesis intends to investigate means of predicting key channel performance parameters such as return loss and insertion loss in high data rate Ethernet cabling standardization using the 40GBASE-T as an example. Insertion loss is the loss of signal power from the transmitting to the receiving ends of a cable which can be caused by factors such as the connectors and physical length of the cable used [8]. On the other hand, return loss is the loss of signal power caused by impedance variations in the structure of a cable or parts associated with it that makes it to reflect back to the source [9]. The research will use the 40GBASE-T as an example to provide guided performance parameter predictions for the proposed channel configurations using the scattering parameters. The prediction method to be presented is equally applicable to ongoing and future high data rate Ethernet cabling standardization such as the 2.5/5GBASE-T and 50/100GBASE-T.

The growing demand for internet of things (IOT) services also makes it imperative to have a reliable Ethernet driven communication network to support the required infrastructure. This is also due to the increasing population in cities that requires adequate provision of services and infrastructure through the use of information and communication technologies (ICTs) to meet the objectives of smart cities development [10] [11]. Therefore, the availability of counterfeit and non-standards compliant twisted pair cables in the market to support the existing Ethernet technologies is also of serious concern to the networking world [12] [13]. This is due to a significant amount of communications cables containing copper clad aluminum cable or other non-standards compliant conductors disguised as category 5e or 6 cables [12], [14], [15].

The use of non-compliant cables can pose serious problems to company networks, the contractors or the installers. For example, the contractors or installers can be charged with criminal liabilities which can be in form of fines, replacement costs and imprisonment [16], [12]. Methods that can be used to identify these sub-standard products are therefore needed [12]. The method to be adopted in this research will be to evaluate the resilience or otherwise of cables to 'handling' stress which is the basic process they can be subjected during installation or reuse. The research presented in this thesis will assess Ethernet cables resilience or otherwise through key performance measurements using the Feature Selective Validation (FSV) method [17] and the Kolmogorov-Smirnov (KS) test [18].

Another problem is to evaluate the impedance profiles or physical integrity of Ethernet cables as a performance indicator before deployment. In the aforementioned situation, where only simple (magnitude) tests are available and time domain tests are inaccessible: as might be the case when historical data only is available or where facilities are not available for reflectometry measurement, a technique to reverse engineer the impedance profile from available S-parameters measurements is needed. This research will therefore provide a method of obtaining the impedance profiles of UTP cables from measured return loss data using genetic algorithms (GA). The genetic algorithm (GA) was selected due to its robustness, ability to manipulate genetic operators in order to mix good features from different solutions to get the best solutions and the use of a population of solutions rather than single solution for searching [19], [20].

Crosstalk, the unwanted signal interference between two wires in the same bundle caused by electrical energy has been found to be one of the major sources of signal degradation in communication systems [21] [22]. The need for a method of predicting crosstalk in twisted pair cables led to the formulation of the standard simple NEXT and FEXT models [23]. However, the inability of the standard simple crosstalk models to predict the peaks and dips often found in measurements led to the derivation of the advanced far-end crosstalk (FEXT) model presented in [22]. This research will therefore evaluate and predict NEXT adapting the "advanced" FEXT model approach presented in [21], [22]. The results of the evaluation will be used to provide improved crosstalk parameters for NEXT simulation and prediction in UTP using Category 6 cables as an example. The research will use the measured NEXT of three Category 6 UTP cables from different manufacturers for evaluation and validation. The evaluation and modeling method can thus be useful to engineers investigating NEXT in the design of data communication systems.

1.2 Aims

This thesis intends to provide a set of analytical and forensic tools for data cabling, with a particular focus on Ethernet cabling to assist designers and those involved in deployment in analyzing cable performance and the reasons behind the actual performance obtained. This is achieved with the following aims:

1. Provide evaluation and modeling method that can help cable engineers investigating crosstalk in the design of data communication systems.
2. Provide cable measurement assessment techniques that can help cable professionals, contractors, installers etc. in the objective selection of cables using the Feature Selective Validation (FSV) method and Kolmogorov-Smirnov (KS) test.
3. Reverse engineer impedance profiles from S-parameters measure-

ments using an optimization technique. The reverse engineering of impedance profiles from S-parameter measurements will find use in situations where only simple (magnitude) tests are available and time domain tests are inaccessible or where facilities are not available for time domain reflectometry measurements.

1.3 Objectives

The aims of the research will be achieved through the following objectives:

1. An assessment of key performance measurements such as return loss and impedance profile of UTP cables using the FSV method and the KS test. The resilience or otherwise of four Category 6 UTP cables from different manufacturers will be evaluated by subjecting them to a series of handling stress. One of the cables to be assessed will be a copper clad aluminum (CCA) cable, widely recognized as being a poor technical solution. These tests will show how the assessment method presented in this thesis can help cable engineers make objective decisions in the selection of cables.
2. Crosstalk evaluation and modeling of unshielded twisted pair (UTP) cables using the Category 6 cables as an example. This is to investigate and provide crosstalk parameters that can help in fast crosstalk prediction in Category 6 UTP cables using the advanced NEXT model. In this research, the NEXT measurements of three Category 6 UTP cables from different manufacturers will be evaluated using the standard simple and advanced models. The NEXT evaluation and modeling method to be presented in this thesis will serve as a useful tool to cable engineers investigating NEXT in data communication systems.
3. Reverse engineer Ethernet cables impedance profile from S-parameters measurements using the return loss data of four Category 6 UTP cables as an example. The technique involves the use of cascaded scattering S-parameters and simple GA to optimize model parameters and extract the impedance profile. This technique of reverse engineering of impedance profile from S-parameters measurements will find application where only simple (magnitude) tests are available and time domain tests are inaccessible or where 'legacy' frequency domain data is the only thing available as reference results and no facilities for time domain reflectometry measurements.

4. Guided simulation to predict the return loss of Ethernet channel configurations based on specifications using impedance variations. The 40GBASE-T will be used as a sample to predict the return loss using specified impedance variations. The guided simulation method is equally applicable to ongoing and future high data cabling and standardization such as 2.5/5 GBASE-T and 50/100GBASE-T.

1.4 Contributions to Knowledge

1. This thesis provides a technique that can be used to evaluate key performance measurements before deployment using the FSV method and KS test that does not rely heavily on human subjective judgement and enables objective decisions in the choice of cables. The method can also help monitor changes in key performance parameters and overcome the difficulties of doing this by eye. The presented assessment method can thus help cable engineers/installers minimize the risk of fines, replacement costs and imprisonment, associated with using counterfeit and non-standards compliant Ethernet cables.
2. It presents a method of evaluating and modeling NEXT in UTP cables using the Category 6 cables as an example. The results of the evaluation was used to provide crosstalk parameters that can help in fast NEXT prediction in Category 6 UTP cables using the advanced NEXT model. The evaluation and modeling method can be used by cable engineers investigating crosstalk in the design of data communication systems.
3. A technique that can be used to reverse engineer impedance profiles of Ethernet cables from return loss measurements using genetic algorithms. This method is very important where time domain tests are inaccessible and only simple (magnitude) tests in the frequency domain is available: as might be the case when historical data only is available or where facilities are not available for reflectometry measurements. The method can thus be used in the aforementioned situations when there is the need to evaluate the impedance profiles or physical integrity of Ethernet cables as a performance indicator before deployment.
4. Provided guided simulation technique that can be used to predict channel performance parameters such as return loss based on specifications in Ethernet cabling standardization research. The method can be used to predict the return loss of Ethernet channels configurations using specified impedance variations. The presented guided

simulation method using the 40GBASE-T as an example will find use in ongoing and future high data cabling and standardization such as 2.5/5 GBASE-T and 50/100GBASE-T.

1.5 Thesis Organization

Chapter one is the introduction to the research that contains the background and problem statement, the aims and objectives, contributions to knowledge and the thesis organization.

Chapter two is the literature review which provides theoretical background required for the research on scattering parameters, performance parameters prediction in Ethernet channels under standardization, optimization and reverse engineering, validation tools and crosstalk.

Chapter three is the methodology that provides the method used in obtaining the measurements used for the research.

Chapter four is the cable measurements assessment using the FSV method and KS test.

Chapter five deals with crosstalk evaluation and modeling.

Chapter six is the reverse engineering of Ethernet cables impedance profiles from measurements.

Chapter seven presents a method of predicting key channel performance parameters that are applicable to ongoing and future high data rate Ethernet cabling standardization research.

Chapter eight presents the conclusions and future work that contains the research summary, conclusion and future work.

Chapter 2

LITERATURE REVIEW

This chapter offers a theoretical background into scattering performance parameters prediction in Ethernet channels under standardization, optimization and reverse engineering, validation tools and crosstalk.

2.1 Scattering Parameters

This section gives a theoretical background on scattering (S) parameters, transfer (T) parameters and transmission line modeling. This section deals with the relationship between these parameters and how they can be used for transmission line modeling.

2.1.1 Scattering (S) Parameters

Scattering (S) parameters matrices are used to compose models for performing accurate signal-integrity and power integrity simulations [24], [25], [26]. The Scattering parameters (S-parameters) for a two port network is the most commonly used in practical applications and serves as the basic building block for generating the high order matrices of larger networks [27], [28]. The simplified diagram of a two port network showing the S-parameter matrix with the incident and reflected waves is shown in Figure 2.1. The S-parameters are the reflection and transmission coefficients between the incident and reflected waves [29].

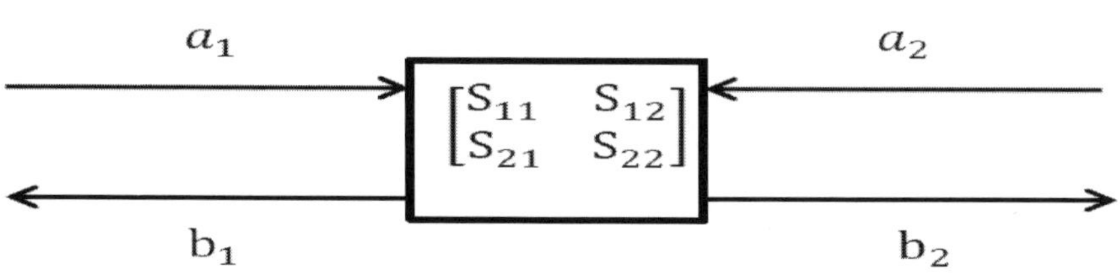

Figure 2.1 Simplified diagram of a two port network showing the scattering parameter matrix

The S-parameter matrix of a two port network which relates the amplitudes of the incident voltage waves (a_1, a_2) to the amplitudes of the reflected voltage waves (b_1, b_2) can therefore be defined as [28], [30]:

$$\begin{bmatrix} b_1 \\ b_2 \end{bmatrix} = \begin{bmatrix} S_{11} & S_{12} \\ S_{21} & S_{22} \end{bmatrix} \begin{bmatrix} a_1 \\ a_2 \end{bmatrix} \tag{1}$$

Equation (1) can be expanded as in [28], [30] to give:

$$b_1 = S_{11}a_1 + S_{12}a_2 \tag{2}$$

$$b_2 = S_{21}a_1 + S_{22}a_2 \tag{3}$$

a_1 and a_2 are the incident signal

b_1 and b_2 are the reflected signal

S_{11} is the input port voltage reflection coefficient

S_{12} is the reverse voltage gain

S_{21} is the forward voltage gain

S_{22} is the output port voltage reflection coefficient

This sub-section has provided the theory of two port S-parameters used in most practical applications. The next section deals with the T-Parameters used for cascading two port networks in transmission line modeling.

2.1.2 Scattering Transfer (T) Parameters

The scattering transfer or transmission (T) parameters are often used in the cascading of two port networks as they allow the convenient representations of networks been cascaded [29], [31]. The T-parameter matrix of a two port network which relates the amplitudes of the waves (a_1, b_1) of the input port to the amplitudes of the waves (a_2, b_2) of the output port can therefore be defined as given in [28], [32] as:

$$\begin{bmatrix} a_1 \\ b_1 \end{bmatrix} = \begin{bmatrix} T_{11} & T_{12} \\ T_{21} & T_{22} \end{bmatrix} \begin{bmatrix} b_2 \\ a_2 \end{bmatrix} \tag{4}$$

Equation (4) can be expanded as in [28] to give:

$$a_1 = T_{11}b_2 + T_{12}a_2 \tag{5}$$

$$b_1 = T_{21}b_2 + T_{22}a_2 \tag{6}$$

2.1.3 Relationship between Scattering (S) and Scattering Transfer (T) Parameters

The relationship between S-parameters and T-parameters are used for the conversion of S-parameters to T-parameters and vice versa when cascading two port networks [31]. The S-parameters can be converted to T-parameters by using equation (7) as given in [31], [33]:

$$S = \begin{bmatrix} \frac{T_{12}}{T_{22}} & \frac{T_{11}T_{22}-T_{12}T_{21}}{T_{22}} \\ \frac{1}{T_{22}} & -\frac{T_{21}}{T_{22}} \end{bmatrix} \tag{7}$$

Similarly, the T-matrix can be converted to S-parameters using equation (8) as expressed in [31], [33].

$$T = \begin{bmatrix} \frac{S_{12}S_{21}-S_{11}S_{22}}{S_{21}} & \frac{S_{11}}{S_{21}} \\ -\frac{S_{22}}{S_{21}} & \frac{1}{S_{21}} \end{bmatrix} \tag{8}$$

The S-parameters from equation (7) are:

$$S_{11}=\frac{T_{12}}{T_{22}},\ S_{21}=\frac{1}{T_{22}},\ S_{12}=\left(\frac{T_{11}T_{22}-T_{12}\,S_{21}}{T_{22}}\right) \text{ and } S_{22}=\left(-\frac{T_{21}}{T_{22}}\right) \tag{9}$$

Similarly, the T-parameters from equation (8) are:

$$T_{11}=\left(\frac{S_{12}S_{21}-S_{11}S_{22}}{S_{21}}\right),\ T_{21}=-\frac{S_{22}}{S_{21}},\ T_{12}=\frac{S_{11}}{S_{21}} \text{ and } T_{22}=\frac{1}{S_{21}} \tag{10}$$

$$\text{Insertion loss} = -20\log_{10}|S_{21}|\,\text{dB} \tag{11}$$

$$\text{Return loss (RL)} = |20.\ |S_{11}||\,\text{dB} \tag{12}$$

This sub-section has provided the relationship between the S-parameters and the T-parameters.

It also deals with how S-parameters can be converted to T-parameters and vice versa when used in the transmission line modeling presented in the next section.

2.1.4 Transmission Line Modeling

Scattering (S) parameters can be defined in terms of the transmission line wave propagation characteristics [29], [34] as:

$$S = \frac{1}{K_s}\begin{bmatrix} (Z_{ck}^{\ 2} - Z_r^{\ 2})\sinh(\gamma_k l) & 2Z_{ck}Z_r \\ 2Z_{ck}Z_r & (Z_{ck}^{\ 2} - Z_r^{\ 2})\sinh(\gamma_k l) \end{bmatrix} \tag{13}$$

where, Z_{ck} and Z_r are the impedance of the cable and reference impedance of the measurement equipment respectively in ohms, l is the length of the cable in meters and γ_k is the propagation constant.

$$K_s = 2Z_{ck}Z_r\cosh(\gamma_k l) + (Z_{ck}^2 + Z_r^2)\sinh(\gamma_k l) \tag{14}$$

$$Z_{ck} = Z_a\left[1 + 0.055\frac{(1-j)}{\sqrt{f_k}}\right] \quad \text{(ohms)} \tag{15}$$

Z_a is the asymptotic impedance of the cable.

The propagation velocity (V_{pk}) is given in [35] as:

$$V_{pk} = \frac{100}{\left(494+\frac{36}{f_k}\right)10^{-9}} \quad \text{(m/sec)} \tag{16}$$

f_k is the frequency at various points i.e. 1Hz, 2Hz, 3Hz etc.

The phase constant(β_k) is given in [35] as:

$$\beta_k = \frac{2\pi f_k 10^6}{V_{pk}} \quad \text{(Radian/m)} \tag{17}$$

If A_{tk} is the specified attenuation limit for a 100m length of cable, then the attenuation constant per meter is given in [35] as:

$$\alpha_k = \frac{A_{ttk}/100}{20 \text{ x} \log(e)} \quad \text{(Neper/m)} \tag{18}$$

In summary, the propagation constant (γ_k) can be expressed as:

$$\gamma_k = \alpha_k + \beta_k \tag{19}$$

The transmission line modeling method presented in this section will be applied in the performance parameters prediction in Ethernet channels under standardization.

2.1.5 Reverse Engineering of Cable Measurements

The method used in this thesis for reverse engineering the transmission line impedance profile from measurements is the cascade connection of multiple transmission line structures. The S-parameter model of the single transmission line section is shown in Figure 2.2 [36].

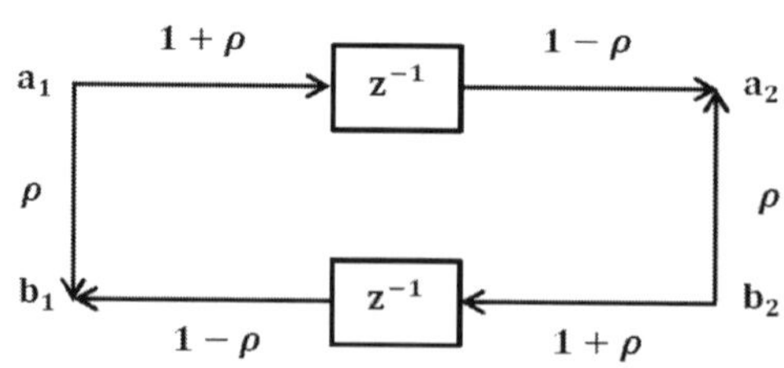

Figure 2.2 Diagram of a single transmission line section

The S-parameters (S_L) of an ideal transmission line model of Figure 2.2 is given in equations (20) and (21) as presented in [36], [37]

$$S_L = \begin{pmatrix} S_{L11} & S_{L12} \\ S_{L21} & S_{L22} \end{pmatrix} \tag{20}$$

S_{11} is the input reflection coefficient of the ideal transmission line

S_{12} is the reverse transmission coefficient of the ideal transmission line

S_{21} is the forward transmission coefficient of the ideal transmission line

S_{22} is the output reflection coefficient of the ideal transmission line

$$S_L = \begin{pmatrix} \rho(1-z^{-2}) & (1-\rho^2)z^{-1} \\ (1-\rho^2)z^{-1} & \rho\,(1-z^{-2}) \end{pmatrix} \frac{1}{1-z^{-2}\rho^2} \tag{21}$$

where, the reflection coefficient $\rho = \frac{Z_c - Z_o}{Z_c + Z_o}$,Z_c+Z_o is the impedance of the cable in ohms, Z_o is the reference impedance.

$z = e^{j2\pi fT}$, where, f *is* the frequency in Hz, T is the propagation time in seconds. The S-parameters of the measured system is given in [36] as:

$$S = \begin{pmatrix} S_{11} & S_{12} \\ S_{21} & S_{22} \end{pmatrix} \tag{22}$$

The S-parameters for the ideal transmission line section given by equation (20) is used to de-embed the small section from the measured two-port system S-parameters as in [36], [37] to obtain the two-port S-parameters given in equation (23):

$$S_R = \frac{1}{S_{L22}S_{11} - |S_L|} \cdot \begin{pmatrix} S_{11} - S_{L11} & S_{L21}S_{12} \\ S_{L12}S_{21} & S_{L22}|S| - S_{22}|S_L| \end{pmatrix} \tag{23}$$

where, $|S|$ is the determinant of the matrix in equation (22) and $|S_L|$ is the determinant of the matrix in equation (20).

The aim of this research is to extract the impedance which is the primary parameter profile from the return loss measurements of the cables, therefore, the return loss (S_{11}) from equation (23) as given in [36] is presented in equation (24):

$$S_{R11} = \frac{S_{11} - S_{L11}}{S_{11}S_{L22} + S_{L12}S_{L21} - S_{L11}S_{L22}} \tag{24}$$

Substituting the S_L parameters of equation (21) into equation (24), a new S_{11} for a single transmission line section is obtained and is given in equation (25):

$$S_{11new} = \frac{-S_{11}+S_{11}\rho^2 z^{-2}-\rho z^{-2}+\rho}{\rho^2+S_{11}\rho z^{-2}-S_{11}\rho-z^{-2}} \tag{25}$$

The S_{11new} in equation (25) is the new S-parameter model for a single transmission line using return loss measurement.

This section has provided the transmission line theory necessary for impedance profiles extraction from return loss measurements with the use of an optimization technique discussed in section (2.3).

2.1.6 Physical Primary (RLGC) Parameters of a Transmission Line

This section presents the physical primary parameters of a transmission line: resistance, inductance, conductance and capacitance (RLGC). The effects of handling stress on the R, L, G, C parameters will be evaluated by incorporating the impedance profile measurements due to handling tests as well as cable properties into their computation. The method can help in evaluating the physical integrity of the cables. The schematic diagram of a twisted pair showing the distance between the centers (D) and the diameter (d) is presented in Figure 2.3.

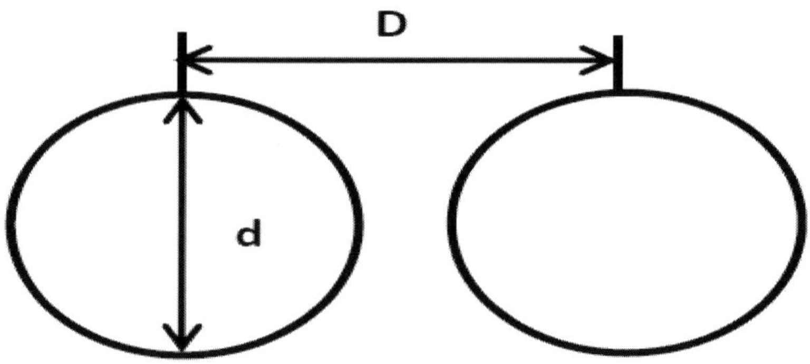

Figure 2.3 Schematic diagram of a twisted pair showing cable dimensions

The general expression for complex propagation constant (γ) is given in [38] as:

$$\gamma = \sqrt{(R + j\omega L)(G + j\omega C)} \tag{26}$$

where R, L, G and C are the primary parameters per unit length of a single pair cable.

They can be computed using their material properties, dimensions etc. as follows [39].

$$R = \frac{2R_s}{\pi d} \qquad (\Omega/m) \qquad (27)$$

The surface resistivity R_s is given in equation (28) as:

$$R_s = \sqrt{\frac{\pi f \mu_c}{\sigma_c}} \qquad (28)$$

$$L = \frac{\mu}{\pi} \ln\left[\left(\frac{D}{d}\right) + \sqrt{\left(\frac{D}{d}\right)^2 - 1}\right] \qquad (H/m) \qquad (29)$$

$$G = \frac{\pi\sigma}{\ln\left[\left(\frac{D}{d}\right) + \sqrt{\left(\frac{D}{d}\right)^2 - 1}\right]} \qquad (S/m) \qquad (30)$$

$$C = \frac{\pi\epsilon}{\ln\left[\left(\frac{D}{d}\right) + \sqrt{\left(\frac{D}{d}\right)^2 - 1}\right]} \qquad (F/m) \qquad (31)$$

where, D is the distance between the centers of the conductors, d is the diameter of each conductor, μ_c is the permeability of the conductor, σ_c is the conductivity of the conductor, f is the frequency in Hz, σ is the conductivity of the insulating or dielectric material, ε is the effective permittivity of the insulating or dielectric material and μ is the permeability of the insulating or dielectric material.

The approximation for the attenuation constant (α) which is the real part of the propagation constant (γ) is given in [38] for high frequency, low-loss transmission lines as:

$$\alpha = \frac{1}{2}\left(R\sqrt{\frac{C}{L}} + G\sqrt{\frac{L}{C}}\right) \qquad (32)$$

Similarly, the approximation for the phase constant (β) which is the imaginary part of the propagation constant (γ) is given in [38] as:

$$\beta = \omega\sqrt{LC} \qquad (33)$$

Therefore, the propagation constant (γ) is:

$$\gamma = \frac{1}{2}\left(R\sqrt{\frac{C}{L}} + G\sqrt{\frac{L}{C}}\right) + j\omega\sqrt{LC} \qquad (34)$$

The R, L, G and C parameters due to the effects of handling stress tests to be carried out can be computed by using the expressions given in [40], [41] as follows:

$$R = \mathrm{Re}(\gamma Z_o) \qquad (\Omega/\mathrm{m}) \tag{35}$$

$$L = \mathrm{Im}(\gamma Z_o)\frac{1}{\omega} \qquad (\mathrm{H/m}) \tag{36}$$

$$G = \mathrm{Re}\left(\frac{\gamma}{Z_o}\right) \qquad (\mathrm{S/m}) \tag{37}$$

$$C = \mathrm{Im}\left(\frac{\gamma}{Z_o}\right)\frac{1}{\omega} \qquad (\mathrm{F/m}) \tag{38}$$

$$\omega = 2\pi f$$

Z_o is the cable impedance measurements in ohms due to handling stress tests.

This section has presented the theory behind the physical primary parameters of a transmission line that will be used in this thesis to evaluate the physical integrity of cables by incorporating the impedance profile measurements due to handling stress into the RLGC calculations as discussed above.

2.2. Parameters Prediction in Ethernet Channels under Standardization

The aim of this section is to investigate means of predicting channel performance parameters such as return loss and insertion loss in high data rate Ethernet cabling standardization research based on specifications. In this thesis, the research considered the 40GBASE-T Ethernet cabling as a case study which is equally applicable to both ongoing and future high data rate Ethernet cabling such as the 2.5/5GBASE-T and 50/100GBASE-T in contribution to the body of knowledge.

2.2.1 Channel Insertion Loss

The 40GBASE-T category 8 cabling is expected to offer a bandwidth level of up to 2000MHz and the channel length is limited to 30m with the use of two connectors [7].

The parameters considered for this research are given in [42], [43] as follows:

$$\text{Channel insertion loss}(IL_{CH}) = 2(IL_1 + IL_2) + IL_3 + IL_4 \tag{39}$$

The 2 in equation (39) is due to the use of two connectors and patch cords.

IL_1 : Insertion loss due to connectors

IL_2 : Insertion loss due to patch cord

IL_3 : Insertion loss due to cable

IL_4 : Insertion loss due to channel deviation where,

$$\left.\begin{array}{c} IL_1 = 0.02 \times \sqrt{f} \text{ (dB)}, 1 \leq f \leq 500 \text{ (MHz)} \\ IL_1 = 0.008 \times \sqrt{f} + 0.00029 \times f + 0.5 \times 10^{-6} \times f^2 \text{ (dB)}, 500 > f \leq 2000 \text{ (MHz)} \\ IL_2 = 1.2 \times IL_3 \\ IL_3 = 1.8 \times \sqrt{f} + 0.005 \times f + \left(0.25/\sqrt{f}\right) \text{ (dB)}, 1 \leq f \leq 2000 \text{ (MHz)} \\ IL_4 = 0.0324 \times \sqrt{f} \text{ (dB)} \end{array}\right\} \quad (40)$$

The performance parameters prediction in Ethernet channels under standardization using the 40GBASE-T as an example as discussed in this section will be achieved with the use of the transmission line modeling method presented in section (2.1.4).

2.3 Optimization

This section discusses the concept of optimization used to reverse engineer the impedance profile of Ethernet cables from S-parameters measurements. The optimization method was applied in this thesis to reverse engineer the impedance profile from return loss measurements of four Category 6 UTP cables. The optimization process presented in thesis will find application where only simple (magnitude) tests are available and time domain tests are inaccessible: as might be the case when historical data only is available or where facilities are not available for reflectometry measurements. The aforementioned situation occurs when there is the need for impedance profiles of cables to evaluate their performance or physical integrity before or after installation.

2.3.1 Optimization and Reverse Engineering

Optimization is the act of obtaining the best solution under given circumstances [44]. Optimization based designs have found use in many disciplines like medicine [45], science [46] and engineering [47]. Optimization is now widely used in many engineering design and decision making processes in which the ultimate goal is to maximize the desired benefit or minimize the effort required [44], [48]. Optimization algorithms have become popular in multi-engineering design problems, mainly due to the availability of high speed computers [49].

Reverse engineering on the other hand, is usually conducted to extract missing ideas, knowledge or design outlook when such information is unavailable [50]. The application of reverse engineering has increased tremendously in various fields due to advances in data analysis and data mining algorithms, coupled with an increase in the availability of cheap computing power [51].

Optimization algorithms have been used in various fields to reverse engineer raw data to obtain desired information [52], [53]. The optimization algorithm chosen for this research is the GA in spite of the availability of other algorithms such as simulated annealing (SA) and particle swarm optimization (PSO). The first reason why the GA was selected for the impedance profiles extraction is the ability to search through large potential solutions as it uses a population based selection, unlike the SA that deals with one individual at each iteration [19]. Another reason why the GA was selected over other optimization techniques is the ability to manipulate genetic operators to mix good features from different solutions to get the required results [19], [20]. The GA is also robust and has also been applied successfully in the optimization and extraction of parameters from measurements using models [20], [54].

This optimization process in this thesis intends to reverse engineer the impedance profile from return loss measurements using the GA. The GA and its working principle are explained in the next section.

2.3.1.1 *Genetic Algorithms*

Genetic algorithms are global stochastic search and optimization algorithms which imitate the process of natural selection and evolution [55]. The GA has been found to be able to search an entire problem space by starting with a number of potential solutions [56], handle a combination of different variables [57] and also extract the desired design parameters [54]. In this thesis, the focus is on using the GA to search the solution that optimizes the S-parameters model and extract the desired parameters (impedance profile) as applied in [58], [59].

The GA starts by generating an initial set of possible solutions called populations. These populations are evolved over a number of iterations to find better solutions depending on the objective function and fitness of the individual. The selected individuals are passed through a mixture of genetic operators (crossover and mutation) before the best individuals after this iteration are selected as solutions to the optimization process [54], [60]. The operations of the GA can be summarized as in the following steps [19], [60]:

(1) Initialization: Generate an initial set of possible solutions (populations).

(2) Fitness calculation: Evaluate the fitness of each possible solution in the population.

(3) Selection: Select individuals from these populations for reproduction to produce new offspring.

(4) Crossover and mutation: Perform crossover and mutation operations to produce new generations (offspring) that combines the characteristics of their parents.

(5) Replace the parents with the new generations.

(6) Go back to step (2) and repeat the procedure until the criteria specified for termination is satisfied.

The GA flowchart for the optimization process is presented in Figure 2.4.

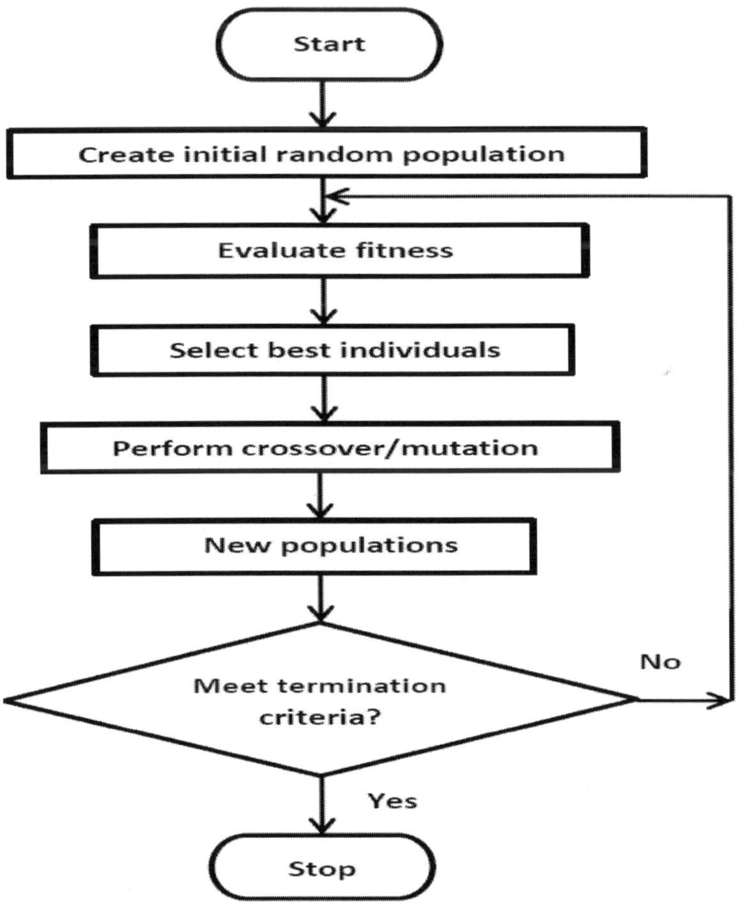

Figure 2.4 Genetic Algorithm flow chart

The three genetic operators used by the GA to get to improve performance results in its operation are explained as follows:

1. **Selection:** The selection (reproduction) process involves choosing two parents from the population for crossing [19]. The chromosomes are selected according to their fitness. Examples of common selection techniques used in the GA are the roulette wheel and tournament. The roulette wheel is one of the most commonly used technique in proportionate selection technique i.e. the selection of individuals based upon their fitness values relative to the fitness of other individuals in the population [19], [61]. This means that the proportion of individual's fitness to the fitness values of the whole population determines the probability of selection of the individual in the next generation [62]. The tournament is another popular selection technique where individuals are chosen at random from the entire population [62], [63]. The individual with the highest fitness value gets selected for the next stage of the GA [62].
2. **Crossover:** The crossover process involves producing new individuals (offspring) from the present individuals (parents) according to a crossover probability [64]. An operation rate of between 0.6 to 1.0 is often used in most problems [61]. If the population is large say 100, a crossover rate of 0.6 is recommended, while for small population say 30, a crossover rate of 0.9 is suggested [61]. Common crossover techniques used in GA are one point, two point and uniform, [65], [66]. In the one point technique, the crossover point is selected randomly within a chromosome [65]. The two parents' chromosomes are then interchanged to produce two new offspring [65]. For the two point technique, two points are selected randomly and the two parent genes are interchanged between these points [67]. In the uniform crossover, the chromosomes are not fragmented for recombination. The gene in each offspring is created by copying it from the parent based on the chosen bit [67].
3. **Mutation:** Mutation is the process of making modifications to a selected individual according to a mutation probability [64]. It is given in [61] that for large population say 100, a mutation rate of 0.001 is recommended, while for small population size say 30, a mutation rate of 0.01 is suggested. Common mutation methods that have been used in optimization problems are uniform and adaptive feasible [68], [69]. In the uniform crossover, the operator changes the value of the chosen gene with a uniform random value selected between the upper and lower bound by the user for that gene [19], [68]. On the other hand, the adaptive feasible mutation randomly

generates directions that are adaptive with respect to the last generation and the feasible region is bounded by constraints [70].

2.3.1.2 *Objective and Fitness Functions*

The objective function is the main source that provides the mechanism for evaluating the status of each chromosome. It is the link between the GA and the system [61]. It can also be simply defined as the system or structure that provides all the design parameters that the GA needs to search for optimal solution. The fitness function on the other hand, determines how well or good is an individual in the current population [56]. The objective function and the corresponding fitness influence the direction of search which makes the GA a useful tool for parameter extraction and optimization [55]. The fitness function determines the performance of individuals in the evolution or iteration process of the GA [56].

2.4 Validation Tools

This section gives an overview of the validation tools used in this thesis report for data comparison or to evaluate the accuracy of predictions. The validation tools considered are the Feature Selective Validation (FSV), Kolmogorov-Smirnov (KS) test and the Mean Absolute Prediction Error (MAPE).

2.4.1 Feature Selective Validation

The FSV was selected for this research to enable the automatic and effective comparison of the data sets from cable measurements and simulations. This is due to the fact that the FSV has been proven to be a robust and helpful technique to quantify visually complex measurement sets, such as those from numerical models, experimental repeatability studies and computational electromagnetics [71], [72]. The FSV can automatically compare and quantify data of any nature by removing the human element of subjectivity and produce objective results needed for a consistent validation in a comprehensible form [71], [73]. The aforementioned characteristic of the FSV have made it the commonly used method for validating and quantifying data [74].

FSV has three indicators that can be used by the user in the analysis of the comparison data. The first is the amplitude difference measure (ADM) which deals with the differences in amplitudes of the data. The second one is the frequency difference measure (FDM) which is measure of the differences in the features of the data compared [71]. The third one is a combination of the ADM and FDM called the global difference measure (GDM) which is a measure of the overall quality of the comparison between the data sets compared [71], [74].

The ADM is calculated as given in [17], [75]:

$$\mathrm{ADM(n)} = \left|\frac{\alpha}{\beta}\right| + \left|\frac{x}{\delta}\right| \exp\left\{\left|\frac{x}{\delta}\right|\right\} \tag{41}$$

where,

$\alpha = (|Lo_1(n)| - |Lo_2(n)|)$

Lo_1 and Lo_2 are the low-pass filter components of the data

$$\beta = \frac{1}{N}\sum_{i=1}^{N}((|Lo_1(i)| + |Lo_2(i)|))$$

$$x = (|DC_1(n)| - |DC_2(n)|)$$

$$\delta = \frac{1}{N}\sum_{i=1}^{N}((|DC_1(i)| + |DC_2(i)|))$$

n is the nth data point, N is the total number of data points, DC_1 and DC_2 are the inverse Fourier transform of the first four data points within the transformed data set for data sets 1 and 2 respectively.

The mean value of the ADM(n) that gives a single figure measure of 'goodness-of-fit' is given in [17] as in equation (41):

$$\mathrm{ADM(n)} = \frac{\sum_{n=1}^{N}\mathrm{ADM(n)}}{N} \tag{42}$$

Similarly, the FDM is obtained from equation (42) as given in [17]:

$$\mathrm{FDM}(f) = 2(|FDM_1(f) + FDM_2(f) + FDM_3(f)|) \tag{43}$$

where,

$$\mathrm{FDM}_1(f) = \frac{|Lo_1{}'(f)| - |Lo_2{}'(f)|}{\frac{2}{N}\sum_{i=1}^{N}((|Lo_1{}'(i)| + |Lo_2{}'(i)|))} \tag{44}$$

$$FDM_2(f) = \frac{|Hi_1{}'(f)| - |Hi_2{}'(f)|}{\frac{6}{N}\sum_{i=1}^{N}((|Hi_1{}'(i)| + |Hi_2{}'(i)|))} \tag{45}$$

$$FDM_3(f) = \frac{|Hi_1{}''(f)| - |Hi_2{}''(f)|}{\frac{7.2}{N}\sum_{i=1}^{N}((|Hi_1{}''(i)| + |Hi_2{}''(i)|))} \tag{46}$$

where, Hi_1 and Hi_2 are the high-pass filter components of the data. The single primes (') are the first derivatives and double prime (") are the second derivatives.

The single figure measure of 'goodness-of-fit' for FDM is calculated the same way as in equation (42).

The GDM is obtained as given in equation (46) as:

$$GDM(f) = \sqrt{ADM(f)^2 + FDM(f)^2} \tag{47}$$

Similarly, the GDM as a single figure 'goodness-of-fit' is found the same way as in equation (42).

The point-by-point comparison of the data set can be represented graphically as (ADMi, FDMi and GDMi), and can also be used to create confidence histograms [74]. These histograms are classified into six quality descriptors: excellent, very good, good, fair, poor and very poor [71].

The average point-by-point comparison of the data sets for evaluating the quality of the results as a single number can be represented as ADMtot, FDMtot and GDMtot [74].

The FSV quantitative and qualitative scale is shown in Figure 2.1 [73], [17].

Table 2.1 FSV quantitative and qualitative Scale

FSV Quantitative Value DM=Difference Measure	**FSV Qualitative Interpretation**
DM < 0.1	Excellent
0.1≥ DM ≤ 0.2	Very good
0.2 ≥ DM ≤ 0.4	Good
0.4 ≥ DM ≤ 0.8	Fair
0.8 ≥ DM ≤ 1.6	Poor
DM > 1.6	Very poor

2.4.2 Kolmogorov-Smirnov (KS) Test

The Kolmogorov-Smirnov (KS) test [18], [76] is non-parametric test that aims to determine if two data sets differ significantly or not. It is a robust test method that has advantage of making no assumption about the distribution of data and is not affected by scale changes [77]. The KS test has been applied in many areas [78], [79], [80] to analyze and compare data. This thesis used the two sample KS test that evaluates the difference between the two cumulative distribution functions (CDFs) of the two data set over the range of "x" in each data set [18], [81].

The KS test uses the P and the test statistic D values to determine whether to reject or accept the null hypothesis. The null hypothesis means the two data set are from the same distribution, while the alternative hypothesis is that they are from different distributions [77], [82].

The test statistic D which is the maximum vertical deviation between the two curves of the CDFs of the two data set is given in [77] as:

$$D_{stat} = \max(|CDF_1(x) - CDF_2(x)|) \tag{48}$$

where,$CDF_1(x)$ is the proportion of values less than or equal to x in the first data set and $CDF_2(x)$ is the proportion of values less than or equal to x in the second data set. The critical value of test statistic D for different significance level can be computed as given in [77], [82] as:

$$D_{crit} = k.\sqrt{\frac{N_1 + N_2}{N_1 . N_2}} \tag{49}$$

Where, N_1 and N_2 is the length of the data sets being compared and value of k can be obtained from [77], [82]. For a confidence level of 95% i.e. significance level (α) of 0.05, the value of k is 1.36. The P values from the KS test also determine if the two data sets differ significantly. The conditions for rejecting or accepting the null hypothesis are [77], [82].

If the test statistic (D_{test}) is greater than the critical value (D_{crit}) when P is less than the significance level (α): null hypothesis is rejected (significant difference between distributions),

If the test statistic (D_{test}) is less than the critical value (D_{crit}) when P is greater than the significance level (α): null hypothesis cannot be rejected or is accepted (no significant difference between distributions). The P value is the main deciding factor in the determination of whether to reject or accept the null hypothesis.

2.4.3 Mean Absolute Prediction Error (MAPE)

The Mean Absolute Prediction Error (MAPE) is used to evaluate the accuracy of the predictions in comparison with experimental data or measurements. It gives the output of the comparison between predictions and measurements as an average percentage error. The MAPE as presented in [83] as:

$$\text{MAPE} = \frac{100}{n}\sum_n \left(\frac{|y_p - y_a|}{y_a}\right) \tag{50}$$

where, y_a are the measured or actual data, y_p are the predicted data and n is the number of data points considered.

2.5 Crosstalk

This section discusses the standard simple and advanced near-end (NEXT) and far-ends (FEXT) crosstalk used in communication systems evaluation and design. Crosstalk is the unwanted signal interference between two wires in the same bundle caused by electrical energy. Crosstalk is one of the major sources of signal degradation in communication systems [84]. Near-end crosstalk is the unwanted signal coupling from the near end of the sending pair to the near end of a receiving pair, while far-end crosstalk is the unwanted signal coupling from the transmitter at the near end into a pair at the far end [85], [86]. The schematic diagram of the near-end and far-end crosstalk is presented in Figure 2.5.

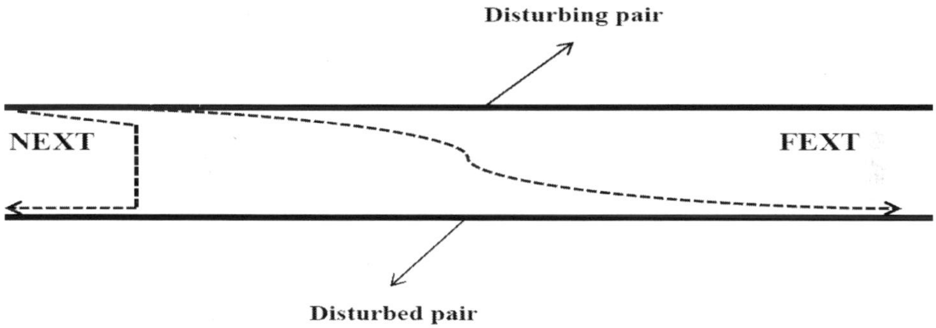

Figure 2.5 Schematic diagram of near end and far end crosstalk

The standard simple FEXT and NEXT models were formulated based on series of measurements for twisted pairs in cable bundles by the Full Service Access Network (FSAN) and American National Standards Institute (ANSI) [23], [87]. However, the standard simple FEXT model has been found not to be too accurate as it could not predict the peaks and dips often found in most crosstalk measurements [21], [22]. The desire to have a better FEXT prediction model than the simple FEXT model led to the evolution of an advanced FEXT model presented in [22]. The advanced FEXT model provided in [22] combined the standard simple FEXT model with some other parameters to get a better prediction model that can imitate the peaks and dips often found in FEXT measurements. Another factor militating against fast crosstalk modeling is that crosstalk constants have been found to be unique (different) for all pair combinations and therefore needs to be decided individually [22], [88]. This research will therefore evaluate and model NEXT using the advanced FEXT method presented in [22] with the aim of determining crosstalk constants for fast NEXT simulation and prediction in UTP cables using Category 6 cables as an example.

2.5.1 Standard Simple Far-end Crosstalk (FEXT) Model

The standard FEXT model is given in [89], [90] as:

$$|\mathrm{H_{FEXT}}(f)|^2 = \mathrm{K_{FEXT}} f^2 l\, |\mathrm{H}(f,l)|^2 \tag{51}$$

The FEXT in dB can be expressed as:

$$\mathrm{FEXT(dB)} = -10\log_{10}|\mathrm{H_{FEXT}}(f)|^2 \tag{52}$$

where, $|\mathrm{H_{FEXT}}(f)|^2$ is the power transmission function of FEXT between two pairs, $|\mathrm{H}(f, l)|$ is the power transmission function of the channel, l is the twisted pair cable length in meters, f is the frequency in hertz (Hz) and $\mathrm{K_{FEXT}}$ is the FEXT crosstalk parameter that determines the prediction limit of the FEXT.

2.5.2 Standard Simple Near-end Crosstalk (NEXT) Model

The standard simple NEXT model is given in [89], [90] as:

$$|\mathrm{H_{NEXT}}(f)|^2 = \mathrm{K_{NEXT}}\, f^{3/2} \tag{53}$$

The NEXT in dB [89] can be expressed as:

$$\mathrm{NEXT(dB)} = -10\log_{10}|\mathrm{H_{NEXT}}(f)|^2 \tag{54}$$

where, $|\mathrm{H_{NEXT}}(f)|^2$ is the transmission function of NEXT between two pair, f is the frequency in hertz Hz and $\mathrm{K_{NEXT}}$ is the NEXT crosstalk parameter that determines the prediction limit of the NEXT.

2.5.3 Advanced Far-end Crosstalk (FEXT) Model

The advanced FEXT model is given in [21], [22] as:

$$|\mathrm{H_{FEXT}}(f)|^2 = |\mathrm{K_{FEXT}} f^2 l\, |\mathrm{H}(f,l)|^2\, \mathrm{K_{NORM}} \left(\cos\frac{2\pi}{5\lambda} - \cos\frac{9\pi}{10\lambda}\right) \tag{55}$$

where, $|\mathrm{H_{FEXT}}(f, l)|^2$ is the FEXT power transfer function between the combination of given pairs, l is the twisted pair cable length in meters and f is the frequency in hertz (Hz), $\mathrm{K_{NORM}}$ is the scaling constant derived from symmetrical pairs resistivity, twisting ratio and diameters.

$\mathrm{K_{NORM}}$ = 3/40 [21], [22] for UTP cables. The cosine expression $(\cos\frac{2\pi}{5\lambda} - \cos\frac{9\pi}{10\lambda})$ is made in such a way that one of them is positive and the other negative so as to maintain the mean value of both cosines nearly zero.

The wavelength is given as $\lambda = \frac{c}{f.\sqrt{\varepsilon_r}}$ (m) (56)

where, f is the frequency in Hz, c is the velocity of light in vacuum and ε_r is the relative permittivity of the cable insulation material.

The FEXT in dB [89] can be expressed as:

$$\text{FEXT(dB)} = -10\log_{10}|H_{FEXT}(f)|^2 \tag{57}$$

2.5.4 Advanced Near-end Crosstalk (NEXT) Model

The method used for the advanced NEXT modeling involves the application of the approach presented in [22] for the advanced FEXT given in equation (55). An examination of the advanced FEXT model in equation (55) indicates that it comprises the standard simple FEXT model in equation (51) and the terms: $\left(\cos\frac{2\pi}{5\lambda} - \cos\frac{9\pi}{10\lambda}\right)$ and K_{NORM}. The research therefore adopted this approach to model the advanced NEXT model by combining the terms: $\left(\cos\frac{2\pi}{5\lambda} - \cos\frac{9\pi}{10\lambda}\right)$ and K_{NORM} with the standard simple NEXT model in equation (53).

Therefore, from the above analysis, the advanced NEXT model is:

$$|H_{NEXT}(f)|^2 = K_{NEXT} f^{3/2} K_{NORM}\left(\cos\frac{2\pi}{5\lambda} - \cos\frac{9\pi}{10\lambda}\right) \tag{58}$$

The NEXT in dB [89] can be expressed as:

$$\text{NEXT(dB)} = -10\log_{10}|H_{NEXT}(f)|^2 \tag{59}$$

The calculation of the average crosstalk constant of each pair of cable (K_{NEXT}) is obtained in [91] and can be expressed as in equation (60) below:

$$K_{NEXT} = 10^{-\left(\frac{A_{NEXT}(f) + 15\log f}{10}\right)} \tag{60}$$

where, $A_{NEXT}(f)$ is the measured crosstalk attenuation and f is the frequency in Hz.

2.5.5 Critical Review of the Crosstalk Models and the Need for Improvements

As already mentioned in section (2.5), the standard simple FEXT and NEXT models were formulated several years ago based on a series of measurements for twisted pairs in cable bundles by FSAN and ANSI [23], [87]. The standard simple FEXT model after evaluation was found not to be too accurate as it could not predict the peaks and dips often found in most measurements [21], [22]. This situation led to the derivation of the advanced FEXT model that combined the standard simple FEXT model with other parameters to obtain a better prediction for UTP cables [22]. The aforementioned scenario necessitated the research on the evaluation of the standard simple NEXT model and the need to adopt the approach in [22] to model NEXT presented in this thesis. The research combined the standard simple NEXT model with other parameters for evaluating and modeling NEXT in UTP cables using Category 6 cables as an example.

Chapter 3

METHODOLOGY

This chapter presents the measurements methodology used for the research. It contains the measurements setup used for obtaining the cable parameters to be used for the research analysis.

3.1 Device and Measurement Setup

DSX-5000 CableAnalyzer [92], [93] can be used for testing and certification of twisted pair cabling such as Category 5e, Category 6, Category 6A cables or Class F_A. The analyzer consists of two modes known as the "main" and "remote" with openings for connections to the standard link interface adapters [94]. The cable to be examined is connected through patch cord plugs to these link adapters and then to the main and the remote for measurements. The cables were tested according to the International Standard ISO/IEC 11801 Class E, T568B pin connection which allows performance of up to 250MHz. The T568 pin connection method was used for the registered jack (RJ45) connection of twisted pair cables before insertion into the patch cords plugs. The DSX 5000 cable tester takes the measurements of the four pairs at the same time. The parameters that can be measured with this tester include insertion loss (attenuation), NEXT, return loss, power sum NEXT (PS-NEXT), HDTDR (High Definition Time Domain Reflectometry) impedance profile across the cable length [94]. The diagram of the cable analyzer measurement setup is shown in Figure 3.1.

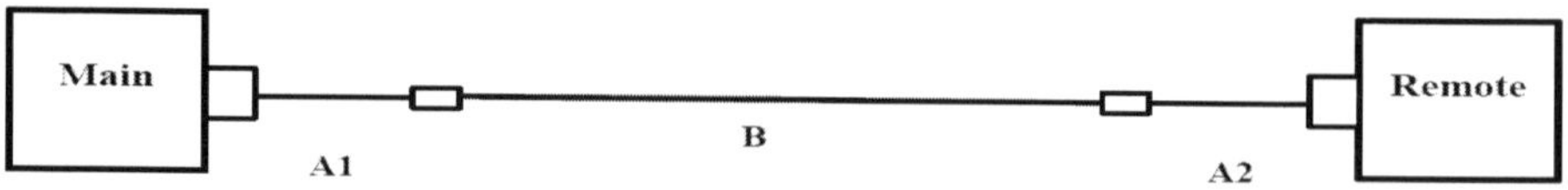

Figure 3.1 The diagram of the cable analyzer measurement set up

A1: Main mode link interface adapter and patch cord plug

A2: Remote mode link interface adapter and patch cord plug

B: Twisted pair cable under test

The photo of the cable analyzer system used for measurements is shown in Figure 3.2.

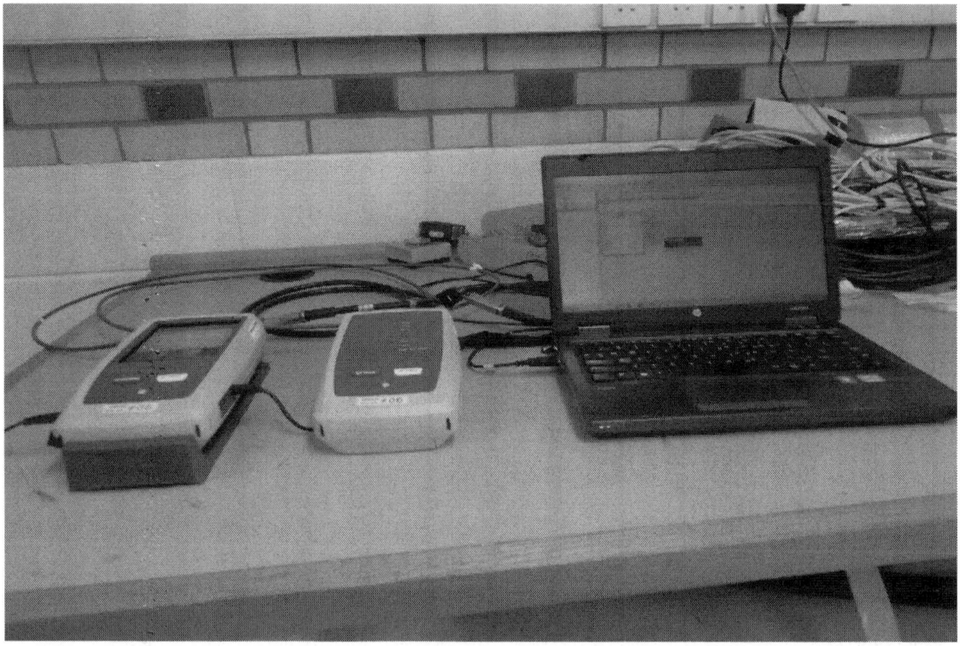

Figure 3.2 The photo of the cable analyzer system used for measurements

3.2 Cables Selected, Pair Numbers, Color Codes and Pin Assignment

Four Category 6 UTP cables from different manufacturers were selected for this research to investigate methods that can be used by cable professionals in making objective decisions in the choice of cables in the market. This is due to a significant amount of communications cables containing copper clad aluminum cable or other nonstandard compliant conductors disguised as category 6 cables [12]. The cables were labelled: cable 1, CCA cable 2 (copper clad aluminum), cable 3 and cable 4. The UTP cables properties and dimensions are:

A. Cable 1

Conductor material: Copper

Insulating material: Polyethylene

Distance between the centers of the conductors (D): 0.99 mm

Diameter of the conductors (d):0.57 mm

B. CCA Cable 2

Conductor material: Aluminum

Cladding material: Copper

Insulating material: Polyethylene

Distance between the centers of the conductors (D):1.03 mm

Diameter of the conductors (d): 0.57 mm

C. Cable 3

Conductor material: Copper

Insulating material: Polyethylene

Distance between the centers of the conductors (D): 0.96 mm

Diameter of the conductors (d):0.54 mm

D. Cable 4

Conductor material: Copper

Insulating material: Polyethylene

Distance between the centers of the conductors (D): 1.01 mm

Diameter of the conductors (d):0.57 mm

The pair numbers, color codes and pin assignment were based on the T568B pin connection used in the cable analyzer. The T568B pin number connection and wire color for the four pair UTP cables are given in [94], [95] and shown in Table 3.1.

Table 3.1 Pin number and wire color of the four pair UTP cable

Pin Number	Wire Color
1	White/orange
2	Orange
3	White/green
4	Blue
5	White/blue
6	Green
7	White/brown
8	Brown

For easy identification and reference purposes in subsequent sections, the pin number, wire color and reference pair color to be used can be expressed as shown in Table 3.2 [94], [95].

Table 3.2 Pin number, wire color and reference pair color for the four pair UTP cable

Pin Number	Wire Color	Reference Pair Color
1,2	White/orange and orange	Orange
3,6	White/green and green	Green
4,5	Blue and white/blue	Blue
7,8	White/brown and brown	Brown

3.3 Measurements due to Handling Stress on Cables

The aim of this research is to provide an assessment method that can be used by cable engineers in the choice of cables. The method adopted in this research will be to evaluate cables resilience or otherwise to handling stress which is the basic process they can be subjected during installation or re-use. Typically, one re-coiling event would happen in many cases during installation, but the cable could be re-coiled up to three times especially when reused [95], [96]. The cables will therefore be subjected to three rounds of coiling-uncoiling operations to mimic handling stress during installation or reuse operations. To present this method of assessment and objective selection of cables, four 30m Category 6 UTP cables from different manufacturers were randomly selected. One of them is a CCA cable, widely perceived as a poor technical solution [12], [15]. The 30m length selected for measurements is due to the maximum length specification for the next generation of cables [7]. The cables are tagged cable 1, CCA cable 2 (copper clad aluminum), cable 3 and cable 4 for easy identification. The cables were coiled to about 30cm diameter which is about the maximum expected in most coiling situations during installation or reuse. The method adopted for the measurement is presented as follows:

Measurement A: UTP cables used to form coils of about 30cm diameter and then stretched out before measurement

Measurement B: UTP cables used for measurements A, reused to form coils of about 30cm diameter and then stretched out before measurement

Measurement C: UTP cables used for measurements B, reused to form coils of about 30cm diameter and then stretched out before measurement

3.4 Impedance Profile Measurements

The impedance profile measurements of the 30m, Category 6 UTP cables were carried out using the procedure explained in section (3.1). However, in this case the DSX-5000 cable tester uses the High-Definition Time

Domain Reflectometry (HDTDR) analyzer embedded in it to measure the impedance profile across the length. The cables measurements methodology is as follows:

Measurement A: UTP cables used to form coils of about 30cm diameter and then stretched out before measurement

Measurement B: UTP cables used for measurements A, reused to form coils of about 30cm diameter and then stretched out before measurement

Measurement C: UTP cables used for measurements B, reused to form coils of about 30cm diameter and then stretched out before measurement

3.5 Measurements Data Processing

The measurements data collected from the procedures presented in sections (3.1) to (3.4) was stored temporarily in the cable analyzer memory during measurements. The measurements from each cable can be easily identified by putting different names on each measurement before collecting and storing them in the analyzer memory. The test results in the memory of the analyzer was transferred to the laptop for extraction by connecting it through a Universal Serial Bus (USB) cable to the 'main' of the analyzer as shown in Figure 3.2. To analyze the test results, the 'LinkWare' management software obtained from the Fluke Network site [92] was installed on the laptop and used to open the test data for analysis.

Chapter 4

CABLES MEASUREMENTS ASSESSMENT

This chapter provides a method of assessing cables measurements using the FSV method and KS tool. The assessment methods presented can help engineers avoid subjective decisions and make objective judgement when selecting cables for network use either for clients or their organizations. An example of where the assessment methods can be applied in decision making is the evaluation of cables resilience or otherwise to 'handling' stress which is the basic process they can be subjected during installation or reuse operations. The cables were subjected to three rounds of coiling-uncoiling measurement tests. However, this is an extreme case test as this does not mean that in every installation, the cables will be subjected to three rounds of coiling-uncoiling operations as used in this research. The procedure involves evaluating key performance parameters measurements such as return loss and impedance profiles of four Category 6 UTP cables using the methods described in chapter three. The four UTP cables were marked as cable 1, CCA cable 2, cable 3 and cable 4 with their material properties and dimensions given in section (3.2).

4.1 Quantifying Variations in Experimental Repeatability

The FSV is used in this section to quantify the variations in experimental repeatability by taking several measurements of return loss and NEXT from which two are selected at random from the repeated cable parameters measurements collected as discussed in section (3.1). The two repeated measurements are known as measurement A and measurement B. The method can be used to verify whether repeatable and sensible measurement practices were taken.

4.1.1 Quantifying Experimental Repeatability using Return Loss Measurement

The comparison of repeated measurements A and B return loss data obtained from the orange pair of cables 1 to 4 is presented in the plots of Figures 4.1 to 4.4. The FSV comparison of measurements A and B return loss data for cables 1 to 4 are shown in Figures 4.5 to 4.8. The y-axis on Figures 4.5 to 4.8 represents the FSV quality, while the x-axis represents the frequency of operation at which the measurements were taken. The results of the FSV return loss comparison for cables 1 to 4 is shown in Table 4.1. An analysis of the FSV GDM_{tot} experimental repeatability results in Table 4.1 shows that cable 1 is 0.0226, CCA cable 2 is 0.0341, cable 3 is 0.0109 and cable 4 is 0.0169. This shows that the similarity between measurements A and B for the four cables is excellent as they all fall below 0.1. This indicates that repeatable and sensible measurements practices were taken.

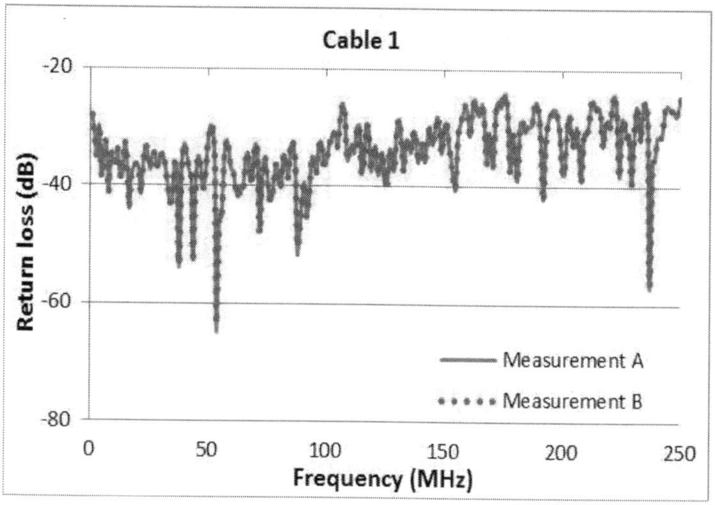

Figure 4.1 Cable 1 return loss comparison

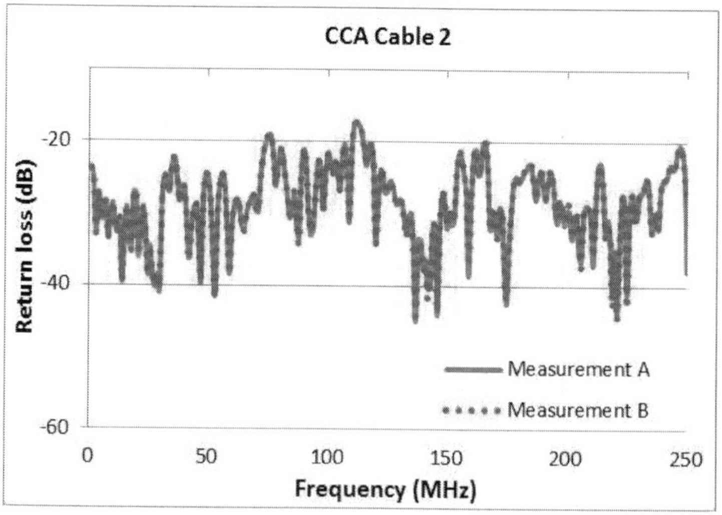

Figure 4.2 CCA Cable 2 return loss comparison

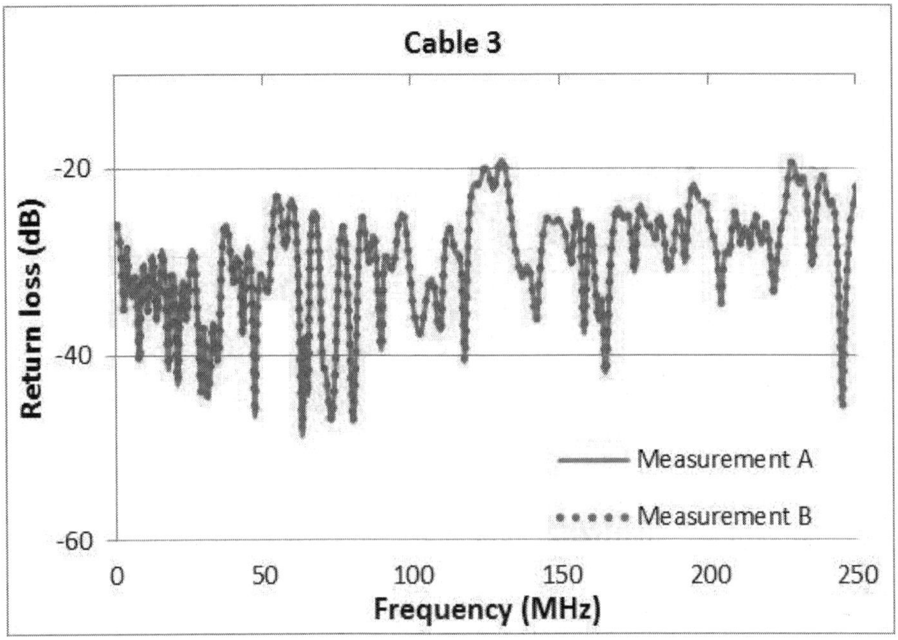

Figure 4.3 Cable 3 return loss comparison

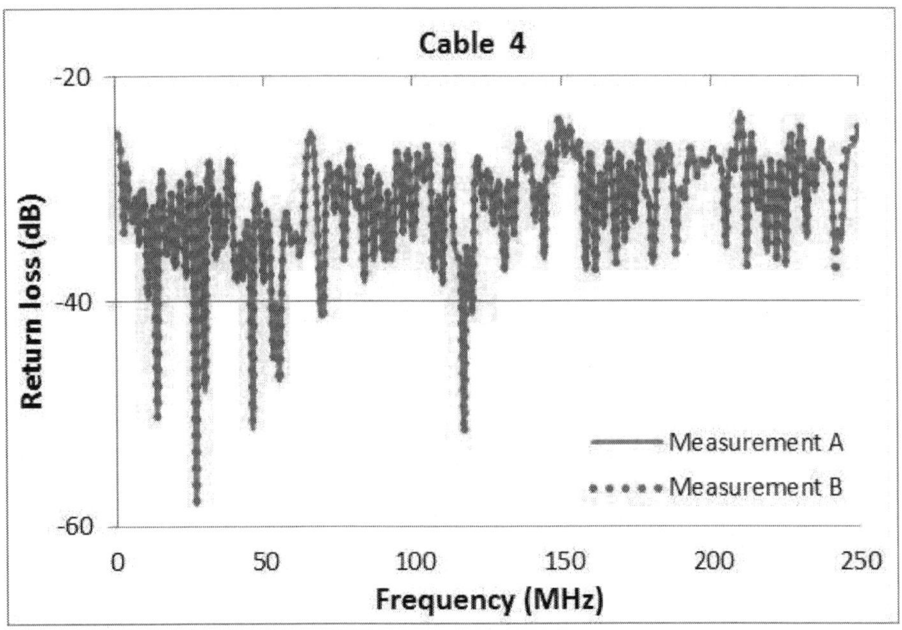

Figure 4.4 Cable 4 return loss comparison

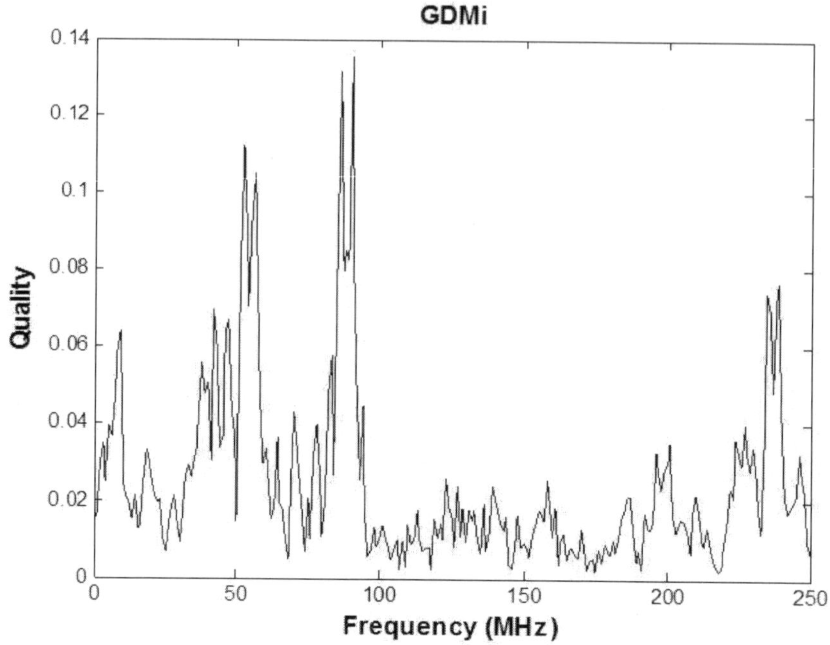

Figure 4.5 Cable 1 return loss comparison using FSV

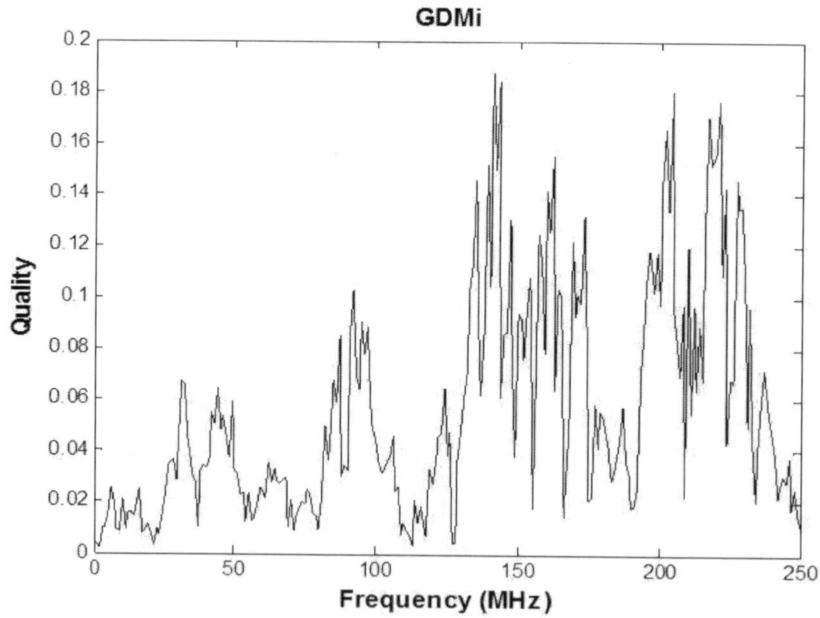

Figure 4.6 CCA Cable 2 return loss comparison using FSV

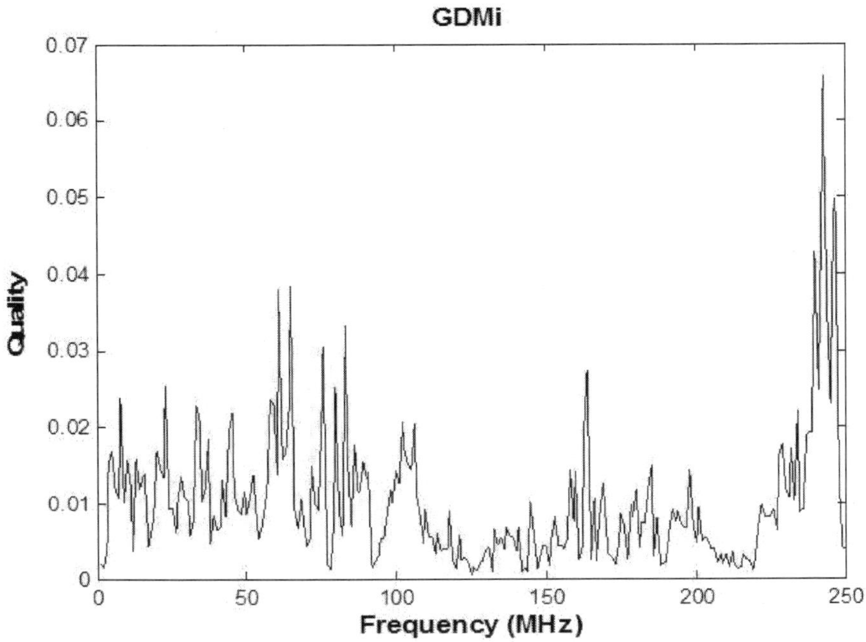

Figure 4.7 Cable 3 return loss comparison using FSV

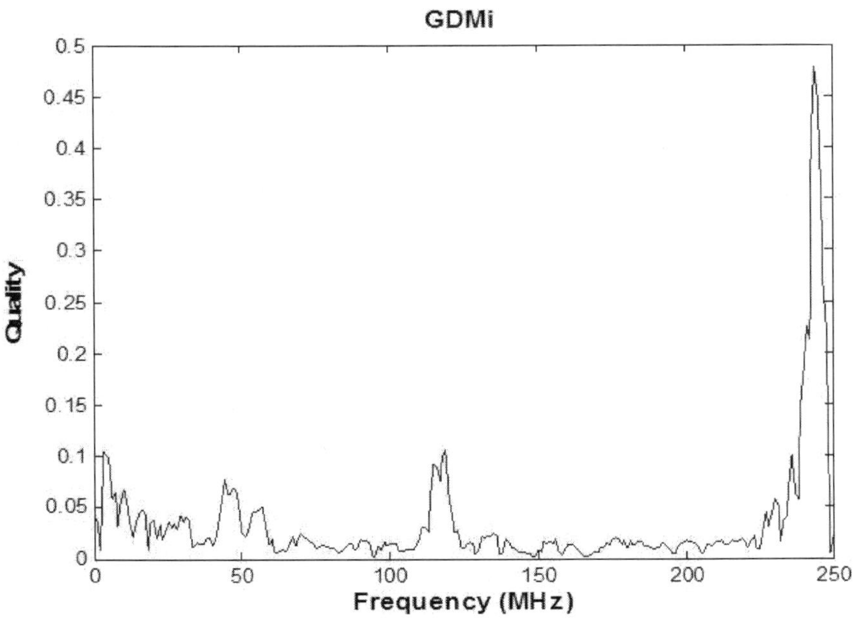

Figure 4.8 Cable 4 return loss comparison using FSV

Table 4.1 Summary of the FSV experimental repeatability results for return loss

	ADM_{tot}	FDM_{tot}	GDM_{tot}
CABLE 1	0.0106	0.0178	0.0226
CCA CABLE 2	0.0198	0.0240	0.0341
CABLE 3	0.0041	0.0093	0.0109
CABLE 4	0.0068	0.0143	0.0169

In summary, the results of the FSV GDM_{tot} comparison in Table 4.1 shows that the similarity between return loss measurements A and B selected at random from repeated measurements of the four cables is excellent using the FSV scale in Table 2.1 as they all fall below 0.1. This indicates that repeatable and sensible measurement practices were undertaken to allow further research analysis.

4.1.2 Quantifying Experimental Repeatability using NEXT Measurement

The NEXT comparison plots for repeated measurements A and B data obtained from the orange pair of cables 1 to 4 are presented in Figures 4.9 to 4.12. The FSV comparison plots for NEXT measurements A and B of cables 1 to 4 are presented in Figures 4.13 to 4.16. The y-axis on Figures 4.13 to 4.16 represents the FSV quality, while the x-axis represents the frequency of operation at which the measurements were taken. The result of the FSV NEXT comparison for cables 1 to 4 is shown in Table 4.2. An analysis of the FSV GDM_{tot} experimental repeatability results in Table 4.2 shows that cable 1 is 0.0463, CCA cable 2 is 0.1415, cable 3 is 0.0182 and cable 4 is 0.0230. This shows that the similarity between measurements A and B for the four cables is excellent as they all fall below 0.1. This indicates that repeatable and sensible measurements practices were taken.

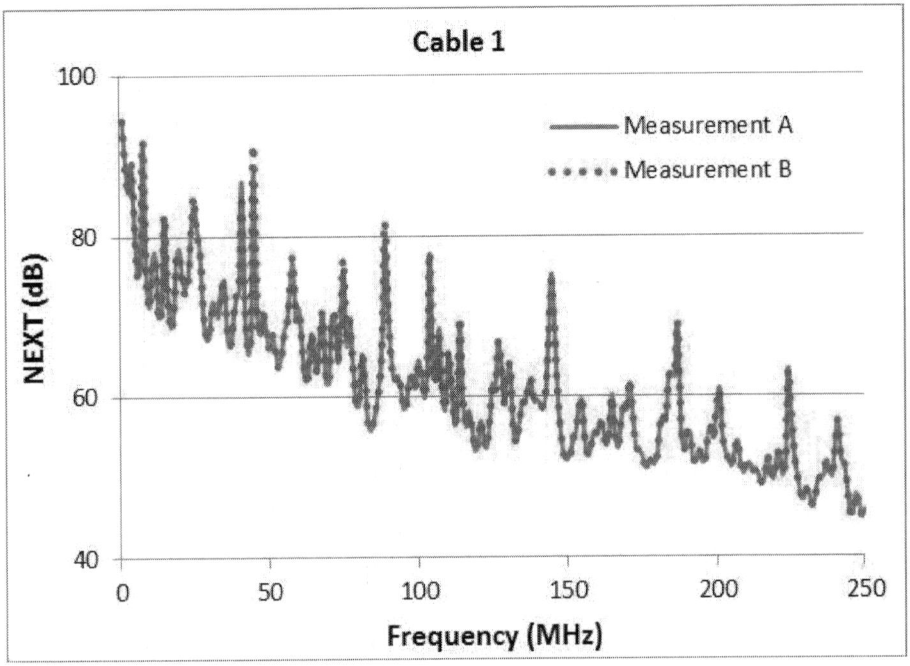

Figure 4.9 Cable 1 NEXT comparison

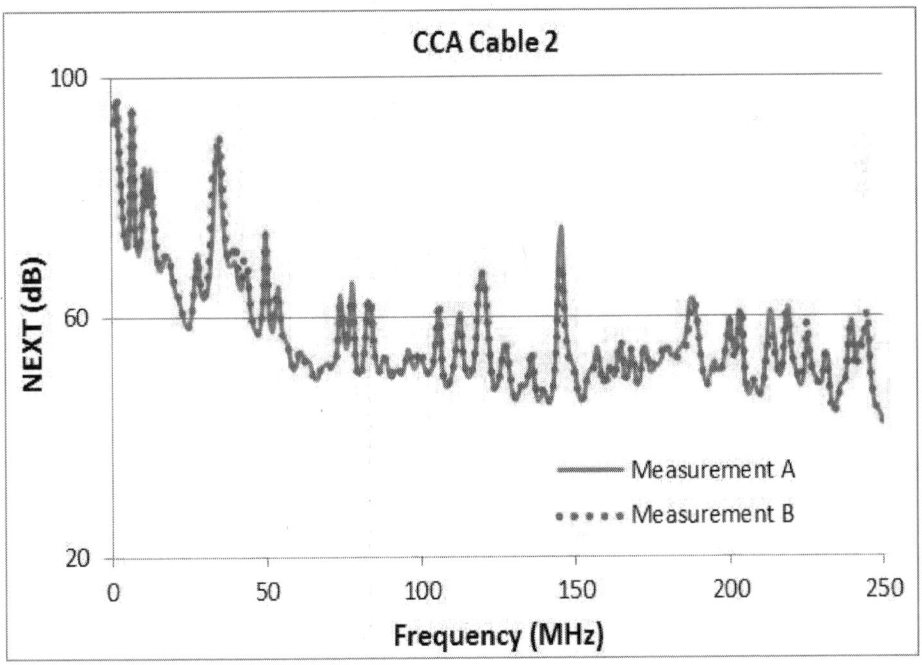

Figure 4.10 CCA Cable 2 NEXT comparison

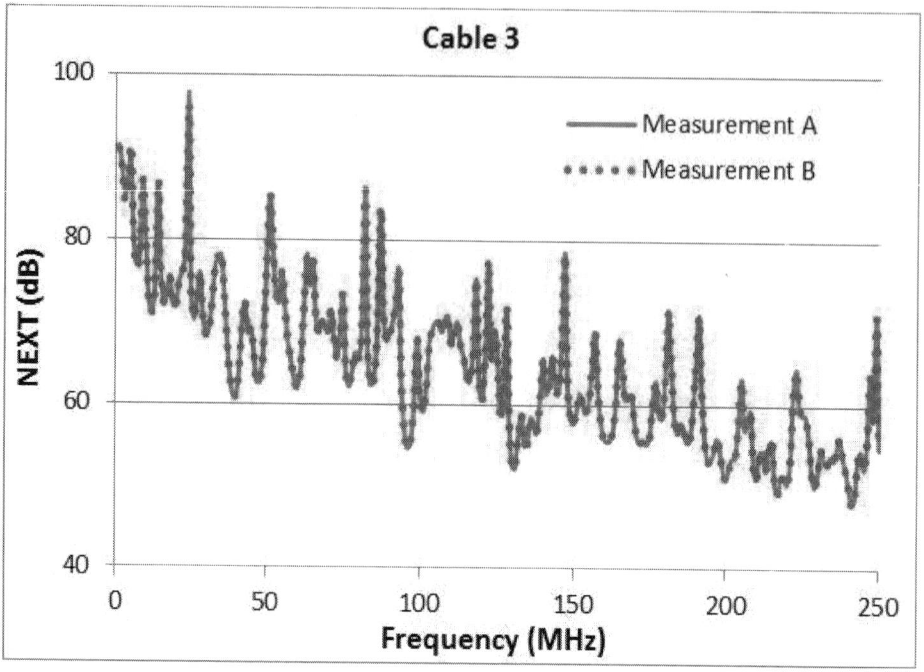

Figure 4.11 Cable 3 NEXT comparison

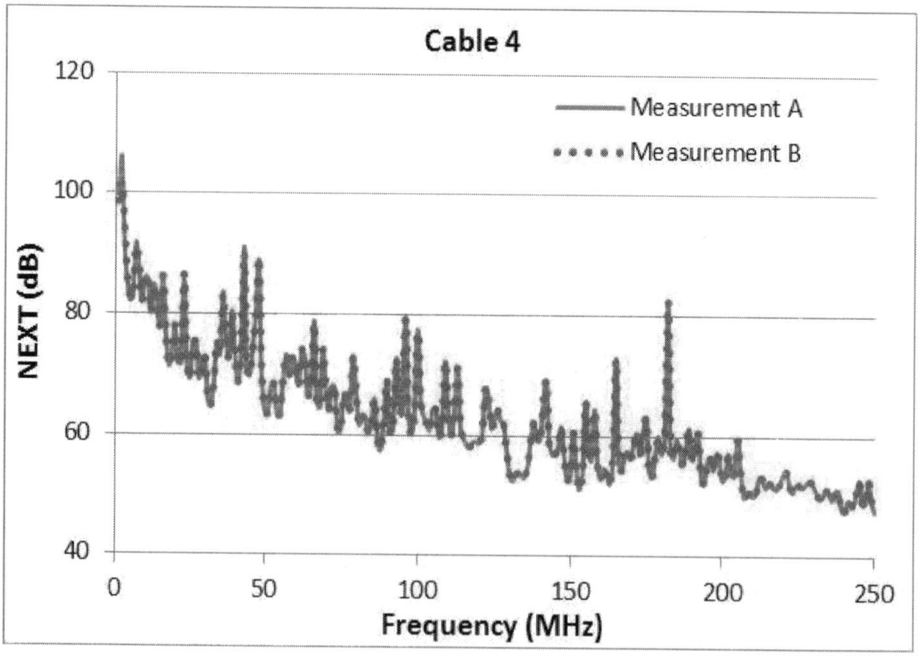

Figure 4.12 Cable 4 NEXT comparison

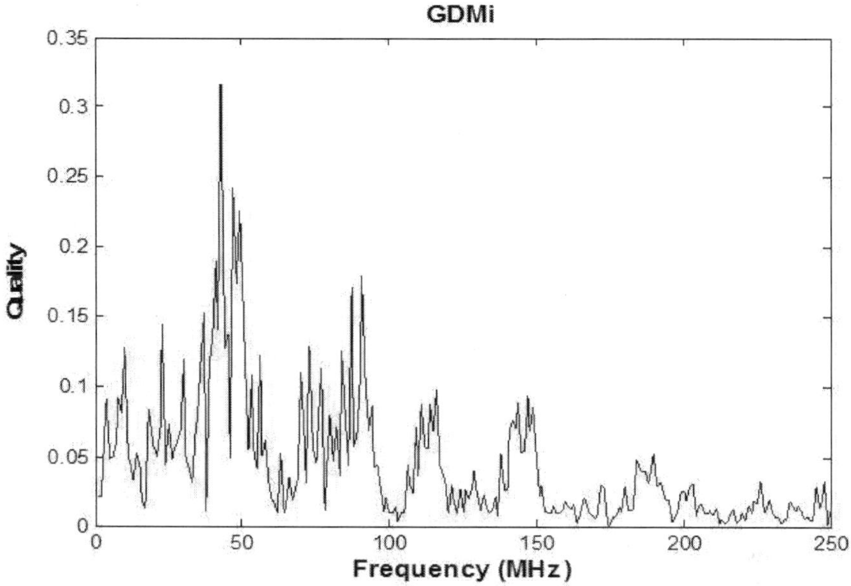

Figure 4.13 Cable 1 NEXT comparison using FSV

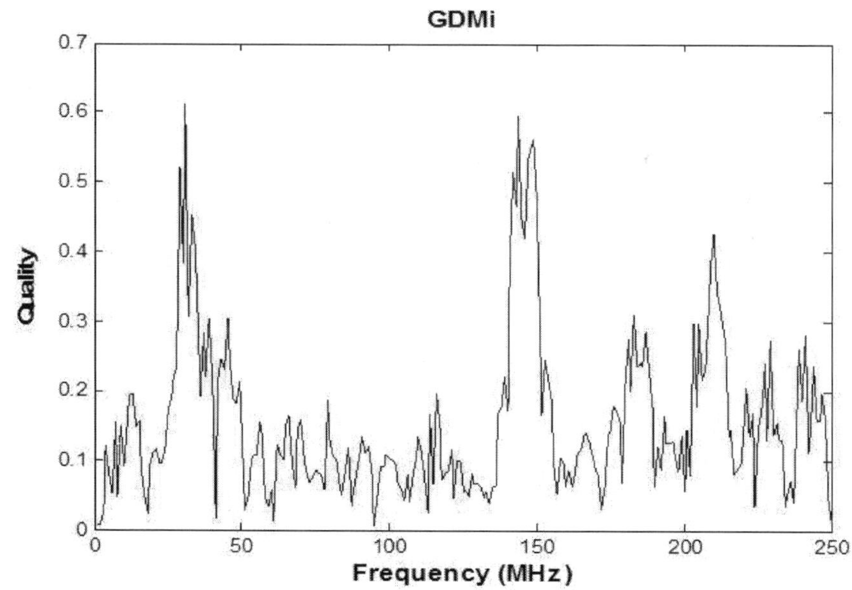

Figure 4.14 CCA Cable 2 NEXT comparison using FSV

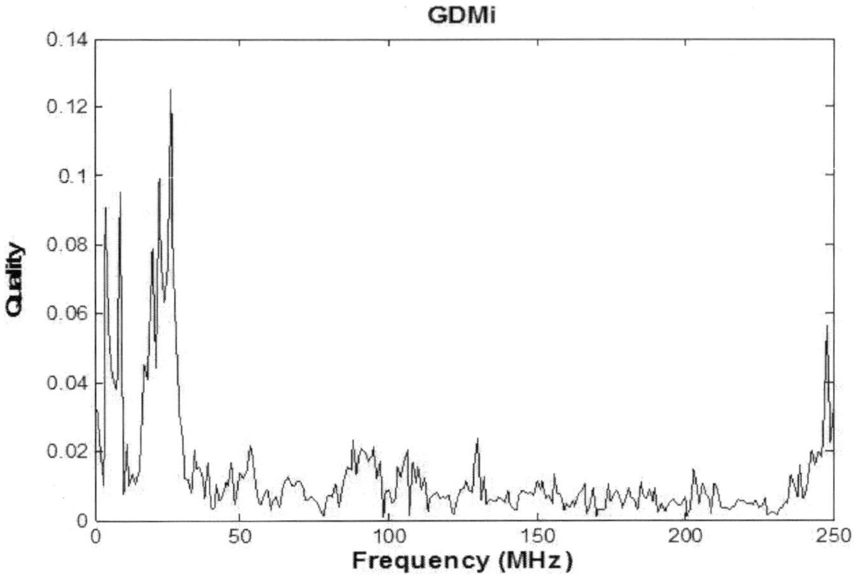

Figure 4.15 Cable 3 NEXT comparison using FSV

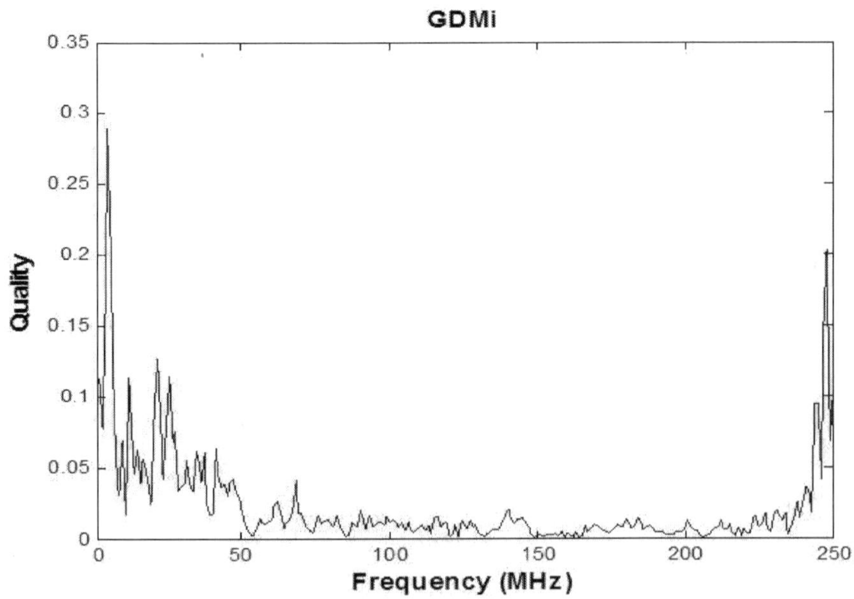

Figure 4.16 Cable 4 NEXT comparison using FSV

Table 4.2 Summary of the FSV experimental repeatability results for NEXT

	ADM_{tot}	FDM_{tot}	GDM_{tot}
CABLE 1	0.0196	0.0380	0.0463
CCA CABLE 2	0.0715	0.1082	0.1415
CABLE 3	0.0058	0.0161	0.0182
CABLE 4	0.0074	0.0203	0.0230

In summary, the results of the FSV GDM_{tot} comparison in Table 4.2 shows that the similarity between NEXT measurements A and B selected at random from repeated measurements of cables 1, 3 and 4 is excellent as they all fall below 0.1, while that for the CCA cable 2 is 0.1415 (approximately 0.1) showing that it is very good using the FSV rating in Table 2.1. This indicates that repeatable and sensible measurement practices were undertaken to allow further research analysis.

4.2 Quantifying Variations in Return Loss Measurements due To Handling Stress

This section presents a method of evaluating cables physical reliability through their resilience or otherwise to 'handling' stress which is the basic process expected during installation or reuse. The method can be used to identify cables with the potential of high susceptibility or otherwise to handling stress expected during installation, particularly reuse operations. The method involves evaluating cables measurements collected through a series of coiling-uncoiling tests to mimic handling stress anticipated during installation or reuse operations. The FSV method and KS test are used to quantify the variations in measurements.

The measurements procedure and terms discussed in section (3.3) is again presented here as follows:

Measurement A: UTP cables used to form coils of about 30cm diameter and then stretched out before measurement

Measurement B: UTP cables used for measurements A, reused to form coils of about 30cm diameter and then stretched out before measurement

Measurement C: UTP cables used for measurements B, reused to form coils of about 30cm diameter and then stretched out before measurement

The comparison plots of the return loss measurements for cables 1 to 4 with Category 6 standard limits [96] are shown in Figures 4.17 to 4.20 using the orange pair. The plots in Figures 4.17 to 4.20 shows that the return loss measurements of cable 1, cable 3 and cable 4 do not cross the Category 6 limits. However, the return loss plots in Figures 4.18 for the CCA cable 2 shows that it crosses the Category 6 limits. This shows that the CCA cable 2 has the potential of been susceptible to degradation if subjected to such stress during installation or reuse.

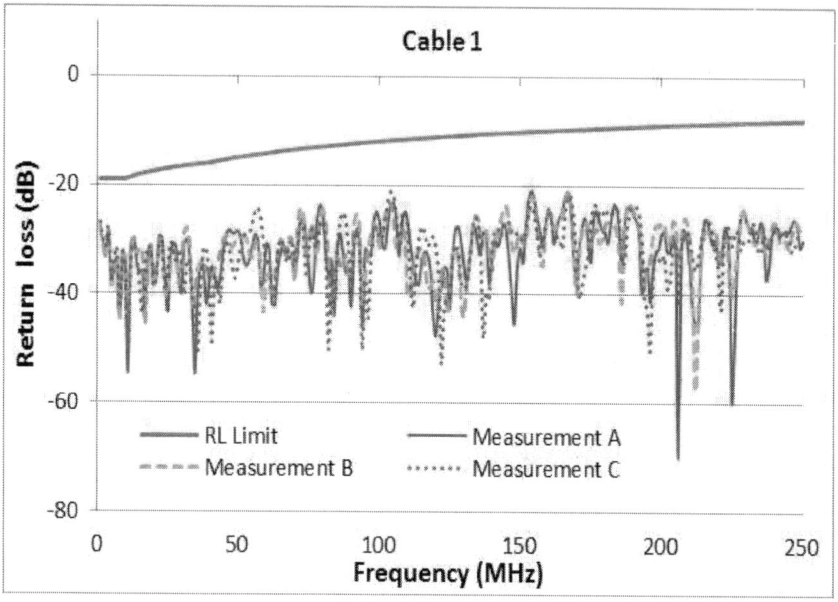

Figure 4.17 Cable 1 return loss measurements

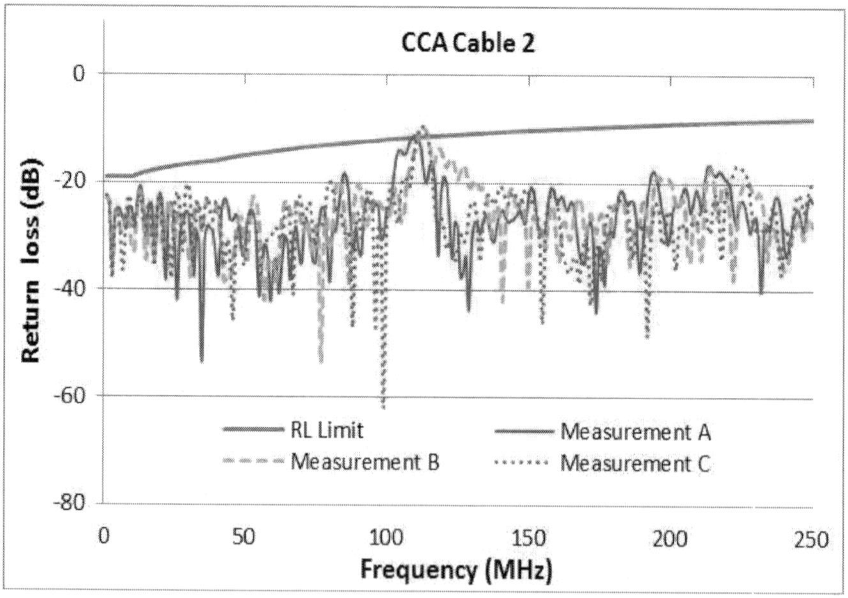

Figure 4.18 CCA Cable 2 return loss measurements

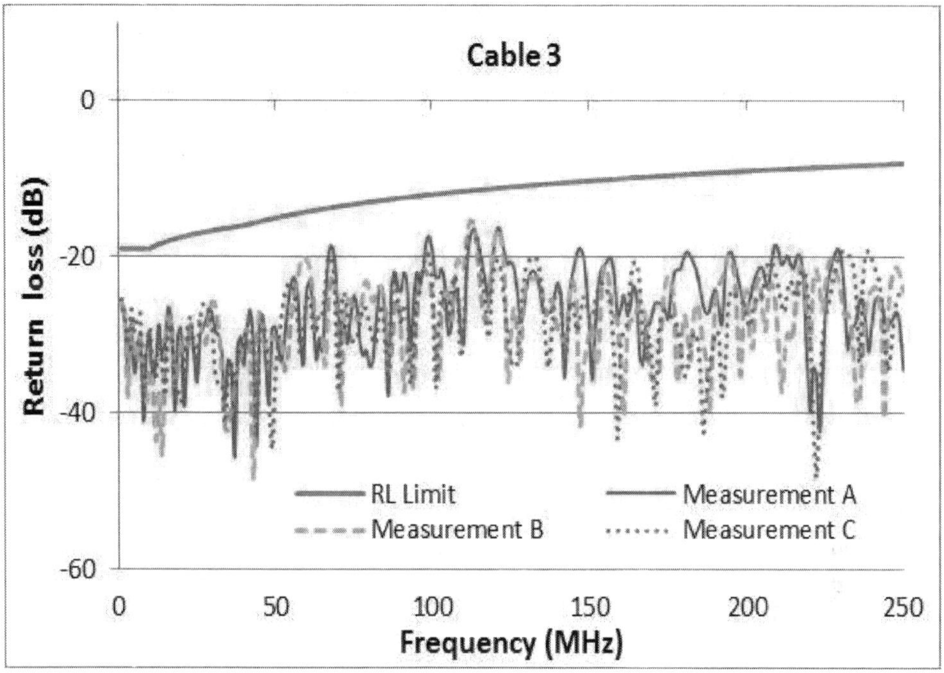

Figure 4.19 Cable 3 return loss measurements

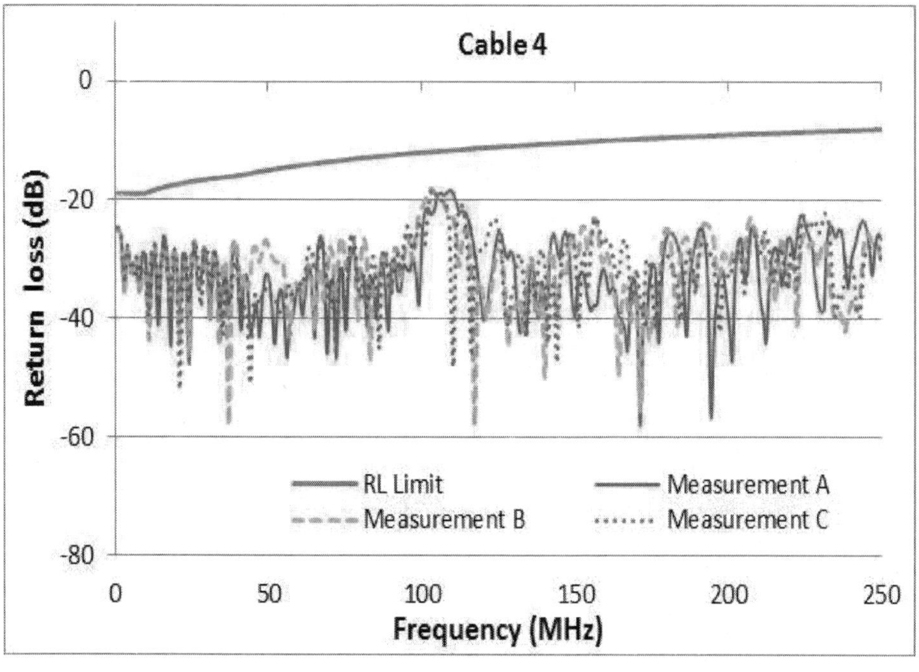

Figure 4.20 Cable 4 return loss measurements

4.2.1 Application of FSV to Quantify Variations in Return Loss Measurement

This section applies the FSV to quantify variations in return loss measurements due to the effects of handling as explained in section (3.3). The aim is to quantify the variations in the return loss between the baseline which is the first test (measurement A) and third test (measurement C) for the four Category 6 cables to be evaluated. The results of the analysis will help identify cables that have high resilience to handling stress with the lowest potential of susceptibility to degradation overtime or reuse. The FSV results of the return loss comparison between measurements A and C of the orange, green, blue and brown pairs for the four cables are presented in Tables 4.3 to 4.6.

Table 4.3 FSV results of the return loss comparison of the orange pair for all the cables

MEASUREMENT A vs C	ADM_{tot}	FDM_{tot}	GDM_{tot}
CABLE 1	0.3279	0.4120	0.5812
CCA CABLE 2	0.3917	0.4576	0.6645
CABLE 3	0.3517	0.3881	0.5832
CABLE 4	0.3790	0.4613	0.6555

Table 4.4 FSV results of the return loss comparison of the green pair for all the cables

MEASUREMENT A vs C	ADM_{tot}	FDM_{tot}	GDM_{tot}
CABLE 1	0.3355	0.4113	0.5903
CCA CABLE 2	0.4374	0.5452	0.7686
CABLE 3	0.3910	0.4290	0.6371
CABLE 4	0.3443	0.3986	0.5854

Table 4.5 FSV results of the return loss comparison of the blue pair for all the cables

MEASUREMENT A vs C	ADM_{tot}	FDM_{tot}	GDM_{tot}
CABLE 1	0.3508	0.4568	0.6418
CCA CABLE 2	0.4244	0.4502	0.6877

CABLE 3	0.3601	0.4628	0.6359
CABLE 4	0.3417	0.3996	0.5844

Table 4.6 FSV results of the return loss comparison of the brown pair for all the cables

MEASUREMENT A vs C	ADM_{tot}	FDM_{tot}	GDM_{tot}
CABLE 1	0.3337	0.4237	0.5998
CCA CABLE 2	0.4239	0.4440	0.6745
CABLE 3	0.3846	0.4349	0.6409
CABLE 4	0.3804	0.4186	0.6265

The FSV GDM results in Tables 4.3 to 4.6 between the baseline (measurement A) and the third test (measurement C) for cable 1 gave the least values of 0.5812 and 0.5998 for the orange and brown pairs respectively, while cable 4 gave the least values of 0.5854 and 0.5844 for the green and blue pairs respectively. The FSV GDM results in Tables 4.3 to 4.6 indicates that cable 1 (orange and brown pairs) and cable 4 (green and blue pairs) gave the least variations between the baseline (measurement A) and the third test (measurement C). The FSV results in Tables 4.3 to 4.6 therefore show that cable 1 and cable 4 gave the highest resilience to the stress the cables were subjected to, after the third test. On the other hand, the FSV GDM results in Tables 4.3 to 4.6 indicates that the CCA cable 2 gave the highest variations between the baseline (measurement A) and the third test (measurement C) for the orange, green, blue and brown pairs with values of 0.6645, 0.7686, 0.6877 and 0.6745 respectively. The FSV results in Tables 4.3 to 4.6 therefore showed that the CCA cable 2 gave the highest susceptibility to the handling stress by providing the highest difference between measurement A and C comparison for all pairs.

The summary of the FSV comparison of return loss measurements A and C for cables 1 to 4 is illustrated with a chart in Figure 4.21.

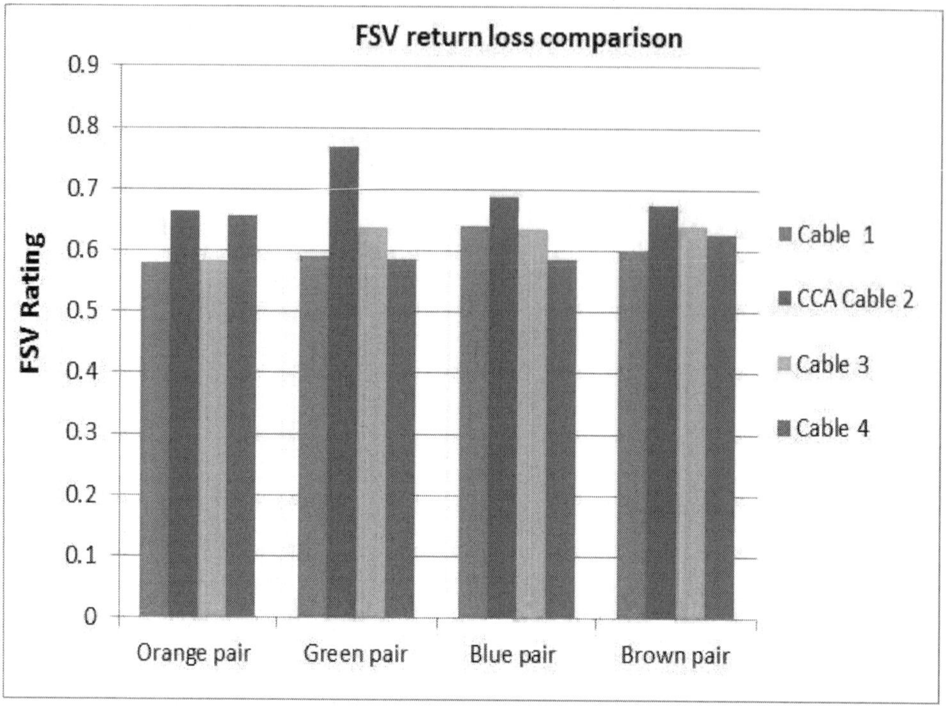

Figure 4.21 FSV return loss comparison chart for the four cables

The summary of the above return loss analysis using the FSV is that cables 1 (orange and brown pairs) and cable 4 (green and blue pairs) gave the least variations between the baseline (measurement A) and the third test (measurement C) as shown in Figure 4.21. On the other hand, the CCA cable 2 (all pairs) gave the highest return loss difference between the baseline (measurement A) and the third test (measurement C) as shown in Figure 4.21 validating their being widely recognized as a poor technical solution or substandard cables for data communication [12], [13]. The FSV results is also corroborated by the return loss plots in Figures 4.17 to 4.20 that showed only the CCA cable 2 crossing the Category 6 limits at some points. However, the FSV results in Table 4.3 to 4.6 shows that the similarity between the baseline (measurement A) and the third test (measurement C) is fair for all the cables pairs. This shows that the impact of the whole length coiling and uncoiling actions after the third test. However, since the three rounds of whole cable length coiling and uncoiling actions is a maximal case test, the cables performance is satisfactory, except for the CCA cable 2. The presented analytical method using the FSV can help engineers/installers make objective decisions in the choice of cables; monitor changes in key performance parameters and enable them overcome the difficulties of doing this by-eye. The KS tool will be used in the next section to determine whether the impact of the three handing stress tests on the cables is significant or not on the return loss.

4.2.2 Application of KS Test to Quantify Variations in Return Loss Measurement

This subsection applies the KS test to determine whether the handling stress test creates a significant difference or not on the return loss comparison between measurement A (baseline) versus measurement C (third test). The KS test already discussed in section (2.4.2) uses the P or the test statistic D values to determine the significance or otherwise of the differences between data sets comparison. However, the P value is the main deciding factor as mentioned in section (2.4.2). The critical value (D_{crit}) to be used as baseline for comparison with the test statistic D values was computed using equation (49) as follows:

$$D_{Critical} = k. \sqrt{\frac{N_1 + N_2}{N_1 . N_2}} \tag{61}$$

N_1 and N_2 are the length of the data sets to be compared; k is 1.36 [77], [82] for a significance value (α) of 0.05 and $N_1 = N_2 = 818$. Therefore, $D_{crit} = 0.067$.

The KS test results for each of the four pairs of cables 1 to 4 is shown in Table 4.7 and Table 4.8 for the test statistic value D and the P value. The conditions for null hypothesis rejection or otherwise is stated below:

If $D > 0.067$ or $P < 0.05$ means null hypothesis is rejected (significant difference)

If $D < 0.067$ or $P > 0.05$ means null hypothesis cannot be rejected or is accepted (no significant difference).

The KS test results in Tables 4.7 and 4.8 indicates that cable 1 (orange pair, blue pair and brown pair), cable 3 (orange pair and green pair) and cable 4 (green pair) showed no significance difference between the baseline (measurement A) and third test (measurement C) comparison as their test P values is greater than 0.05 and D values is less than 0.067. However, the CCA cable 2 (all pairs) showed significant difference between measurement A and measurement C comparison as the P values of all the pairs are lower than 0.05 and D values greater than 0.067.

Table 4.7 D values of the return loss comparison A versus C for the four cables

A vs C D Value	CABLE 1	CCA CABLE 2	CABLE 3	CABLE 4
Orange Pair	0.045	0.099	0.042	0.105
Green Pair	0.075	0.169	0.045	0.037
Blue Pair	0.065	0.084	0.138	0.075
Brown Pair	0.056	0.068	0.082	0.093

Table 4.8 P values of the return loss comparison A versus C for the four cables

A vs C P Value	CABLE 1	CCA CABLE 2	CABLE 3	CABLE 4
Orange Pair	0.329	0.000	0.427	0.000
Green Pair	0.017	0.0001	0.327	0.585
Blue Pair	0.052	0.005	0.0001	0.017
Brown Pair	0.128	0.035	0.006	0.001

The summary of the comparison of the return loss measurements A and C using the KS test for the four cables is illustrated with a chart for the D values in Figure 4.22 and the P values in 4.23.

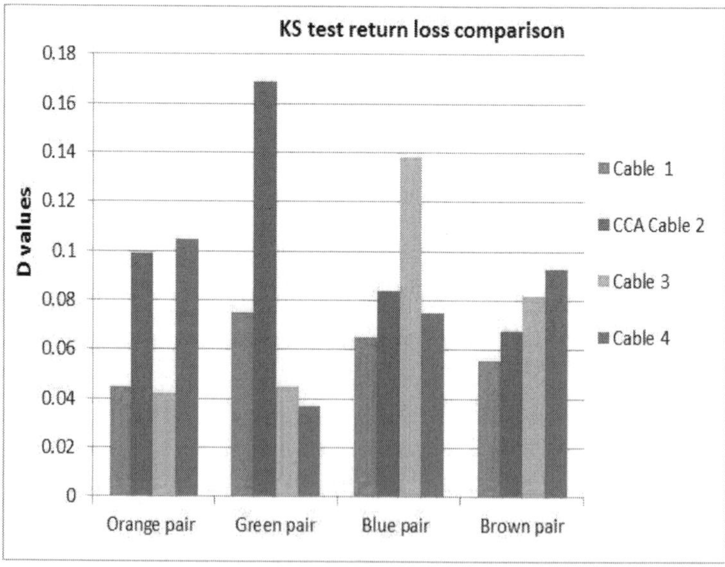

Figure 4.22 KS test D values chart for the return loss comparison of the four cables

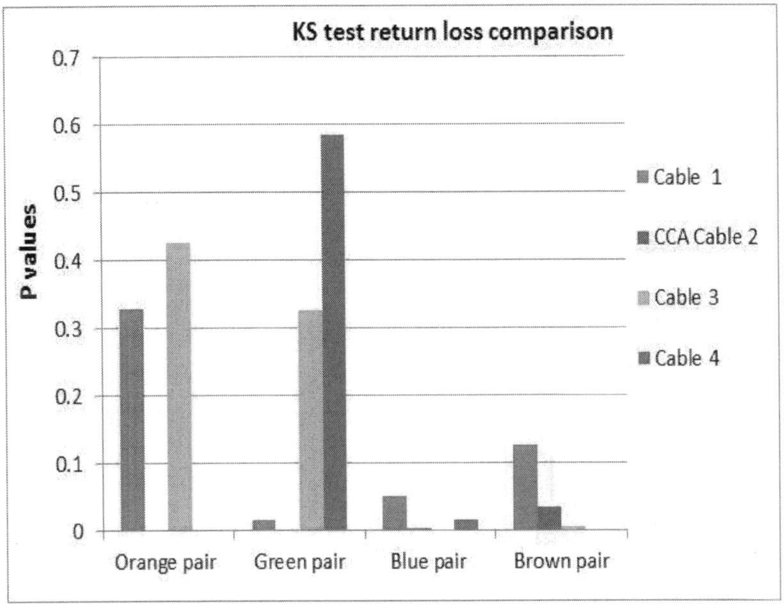

Figure 4.23 KS test P values chart for the return loss comparison of the four cables

The summary of the KS test results for return loss presented in Figures 4.22 and 4.23 is that cable 1 gave the best result between measurement A (first test) and measurement C (third test) comparison as it showed no significant difference in three of it pairs (orange, blue and brown) followed by cable 3 (orange and green pairs) as their P values are greater than 0.05 and D values are less than 0.067. On the other hand, the CCA cable 2 provided the worst results as it gave a significant difference between measurement A and measurement C comparison for all the pairs as their P values are below 0.05 and D values greater than 0.067 as shown in Figure 4.22 and 4.23. The results obtained using the KS test corroborates the results of the FSV tool that the CCA cable 2 gave the worst output, validating their being widely recognized as a poor technical solution or substandard cables for data communication [12], [13].

4.3 Quantifying Variations in Impedance Measurement due to Handling Stress

This section uses the FSV to quantify the variations in impedance profile measurements due to a series of coiling and uncoiling tests used to represent handling stress on the UTP cables. The measurements methodology used for the collection of the impedance profile due to handling stress have been discussed in sections (3.1) and (3.3). Measurements procedure and terms are again presented in this section for easy reference as:

Measurement A: UTP cables used to form coils of about 30cm diameter and then stretched out before measurement

Measurement B: UTP cables used for measurements A, reused to form coils of about 30cm diameter and then stretched out before measurement

Measurement C: UTP cables used for measurements B, reused to form coils of about 30cm diameter and then stretched out before measurement

The impedance profile measurements for cables 1 to 4 using the orange pair are shown in Figures 4.24 to 4.27. A view of the graphs in Figures 4.24 to 4.27 shows that CCA cable 2 and cable 3 plots in Figures 4.25 and 4.26 gave more than 98Ω and 102Ω.

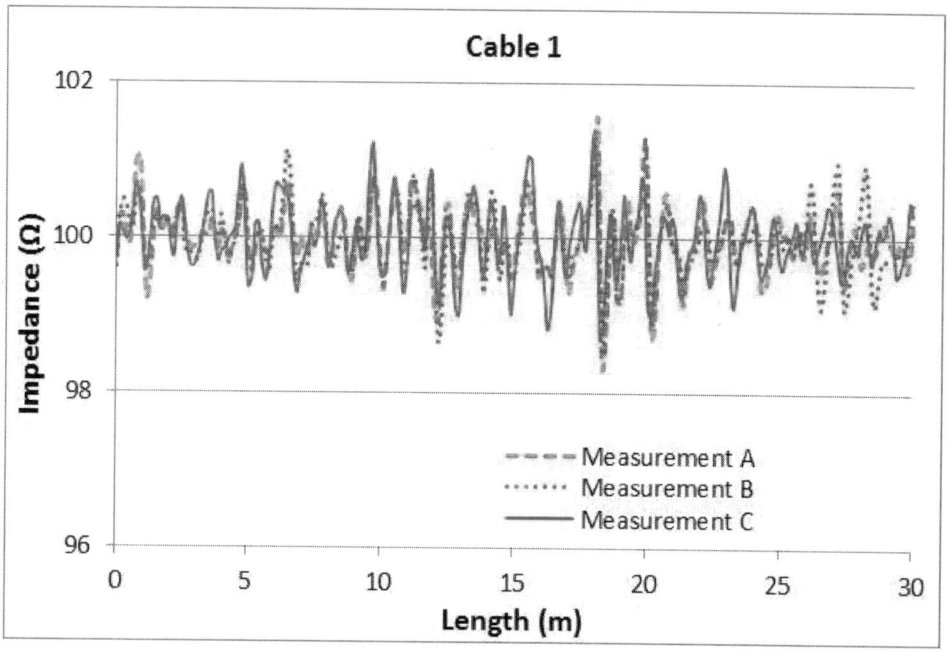

Figure 4.24 Cable 1 impedance profile measurements

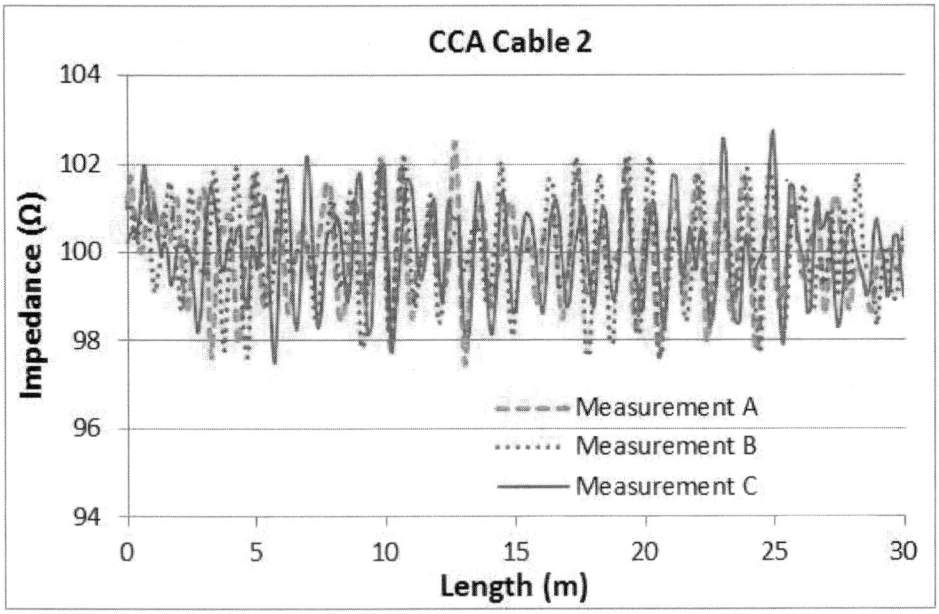

Figure 4.25 CCA Cable 2 impedance profile measurements

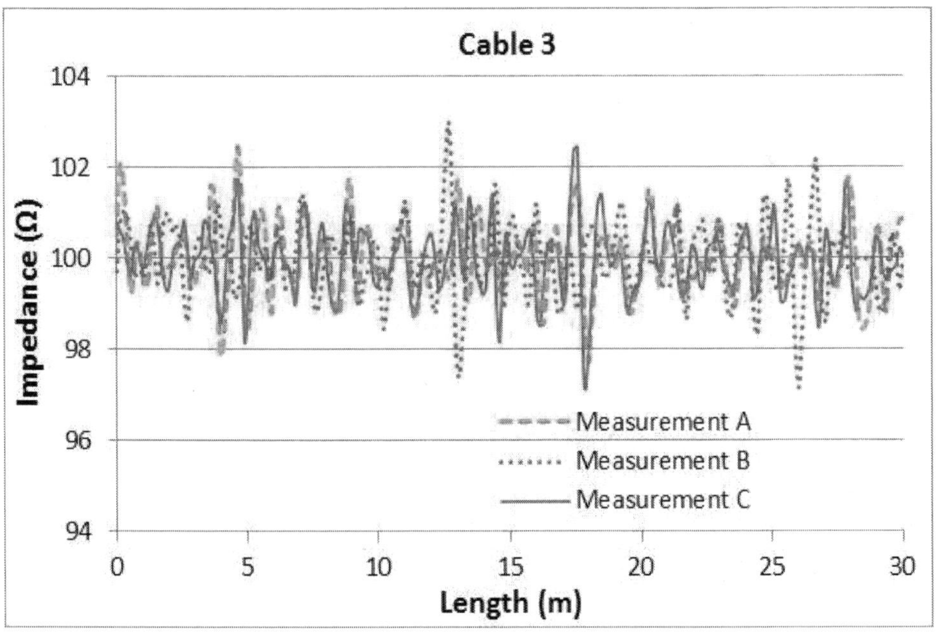

Figure 4.26 Cable 3 impedance profile measurements

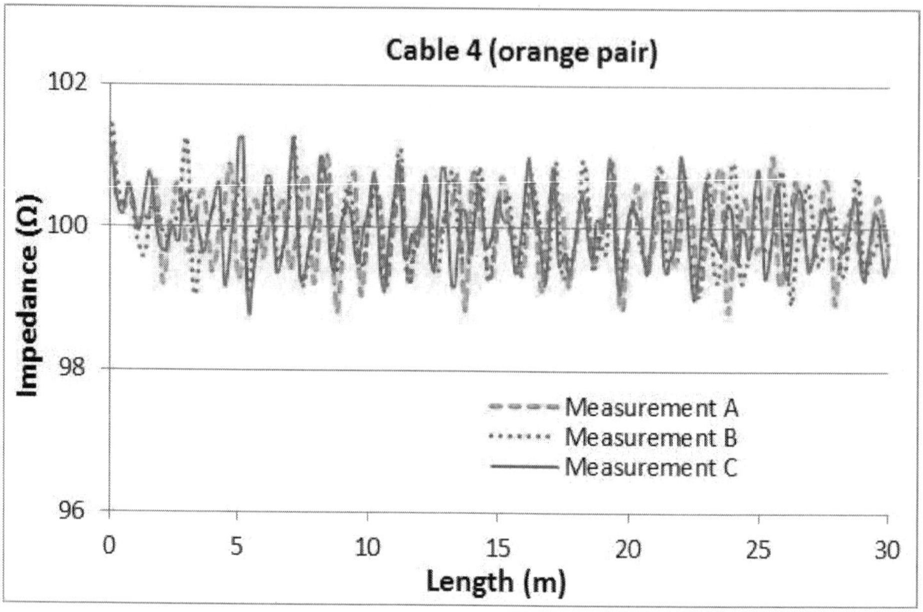

Figure 4.27 Cable 4 impedance profile measurements

4.3.1 Application of FSV to Quantify Variations in Impedance Measurement

This section applies the FSV to quantify the variations in impedance profile measurements A (first test) and measurement C (third test). The FSV results of the impedance profile comparison between measurements A and C of pairs 1 to 4 for the four cables are presented in Tables 4.9 to 4.12. The FSV GDM results in Tables 4.9 to 4.12 between the baseline (measurement A) and the third test (measurement C) for cable 1 gave the least values of 0.5479,0.4969 and 0.5462 for the green, blue and brown pairs respectively, while cable 3 gave the least value of 0.4872 for the orange pair. The FSV results in Tables 4.9 to 4.12 therefore show that cable 1 gave the highest resilience to the stress the cables were subjected after the third test. On the other hand, the FSV GDM results in Tables 4.9 to 4.12 indicates that the CCA cable 2 gave the highest variations between the baseline (measurement A) and the third test (measurement C) for the orange and brown pairs with values of 0.5886 and 0.6408 respectively, while cable 4 gave the highest variations for the green and blue pairs with values of 0.6570 and 0.6864 respectively. The FSV results in Tables 4.9 to 4.12 therefore showed that the CCA cable 2 and cable 4 gave the highest susceptibility to the handling stress by providing the highest difference between measurement A and C comparison for the aforementioned pairs.

Table 4.9 FSV results of the impedance profile comparison of the orange pair for the four cables

MEASUREMENT A vs C	ADM_{tot}	FDM_{tot}	GDM_{tot}
CABLE 1	0.2977	0.3353	0.5101
CCA CABLE 2	0.3493	0.4052	0.5886
CABLE 3	0.2825	0.3314	0.4872
CABLE 4	0.3227	0.3480	0.5287

Table 4.10 FSV results of the impedance profile comparison of the green pair for the four cables

MEASUREMENT A vs C	ADM_{tot}	FDM_{tot}	GDM_{tot}
CABLE 1	0.3103	0.3745	0.5479
CCA CABLE 2	0.3521	0.3790	0.5692
CABLE 3	0.3413	0.3567	0.5490
CABLE 4	0.3897	0.4432	0.6570

Table 4.11 FSV results of the impedance profile comparison of the blue pair for the four cables

MEASUREMENT A vs C	ADM_{tot}	FDM_{tot}	GDM_{tot}
CABLE 1	0.3137	0.3109	0.4969
CCA CABLE 2	0.3211	0.3750	0.5546
CABLE 3	0.3095	0.4080	0.5694
CABLE 4	0.4008	0.4667	0.6864

Table 4.12 FSV results of the impedance profile comparison of the brown pair for the four cables

MEASUREMENT A vs C	ADM_{tot}	FDM_{tot}	GDM_{tot}
CABLE 1	0.3494	0.3359	0.5462
CCA CABLE 2	0.3820	0.4395	0.6408
CABLE 3	0.3321	0.3813	0.5661
CABLE 4	0.3577	0.3934	0.5822

The summary of the FSV results for the comparison between impedance profiles measurements A and C for the four cables is illustrated with a chart in Figure 4.28.

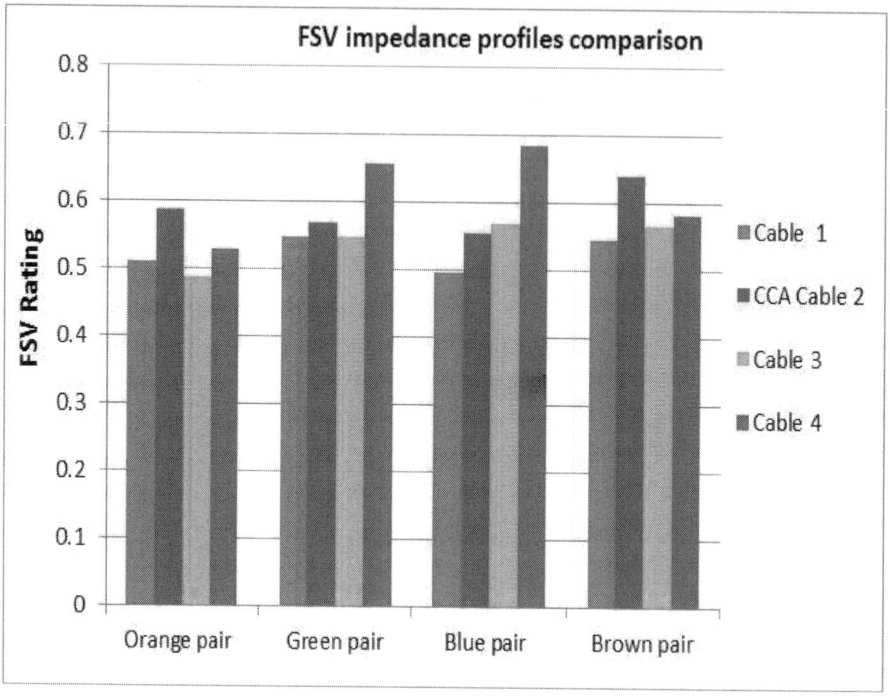

Figure 4.28 FSV impedance profiles comparison chart for the four cables

The summary of the impedance profiles analysis using the FSV is that cable 1 (green, blue and brown pairs) gave the least variations between the baseline (measurement A) and the third test (measurement C) as shown in Figure 4.28. On the other hand, the CCA cable 2 (orange and brown pairs) and cable 4 (green and blue pairs) gave the highest return loss difference between the baseline (measurement A) and the third test (measurement C) as shown in Figure 4.28. However, the FSV results in Tables 4.9 to 4.12 shows that the similarity between the baseline (measurement A) and the third test (measurement C) is fair for all the cables pairs. This shows that the impact of the whole length coiling and uncoiling actions after the third test. However, since the three rounds of whole cable length coiling and uncoiling actions is a maximal case test, the cables physical reliability is adequate. The presented analytical method using the FSV can help engineers/installers make objective decisions in the choice of cables; monitor changes in key performance parameters and enable them overcome the difficulties of doing this by eye. The KS tool will be used in the next section to determine whether the impact of the three handing stress tests on the cables is significant or not on the impedance profile.

4.3.2 Application of KS Test to Quantify Variations in Impedance Measurement

This section applies the KS tool to determine whether the impact of the handling stress tests on the cables is significant or not on the impedance profile comparison between measurement A (baseline) versus measurement C. The critical value (D_{crit}) to be used as baseline for comparison with the test statistic D values was computed using equation (49). In this case, k is 1.36 [77], [82] for a significance value (α) of 0.05, $N_1 = N_2 = 233$ which gives $D_{crit} = 0.1260$ on computation. As stated in section (2.4.2), the null hypothesis cannot be rejected or is accepted if the P values from the KS test are greater than the significance value (α) or the test statistic D values are less than the critical value (D_{crit}) computed for the data sets and vice versa. This can be explained as follows:

If $D > 0.126$ or $P < 0.05$ means null hypothesis is rejected (significant difference)

If $D < 0.126$ or $P > 0.05$ means null hypothesis cannot be rejected or is accepted (no significant difference). The KS tool results for each of the four pairs of cables 1 to 4 is shown in Table 4.13 and Table 4.14 for the test statistic value D and the P value.

Table 4.13 D values of the impedance profile comparison A versus C for the four cables

A vs C D VALUE	CABLE 1	CCA CABLE 2	CABLE 3	CABLE 4
Orange Pair	0.082	0.052	0.064	0.077
Green Pair	0.073	0.069	0.069	0.060
Blue Pair	0.060	0.064	0.086	0.039
Brown Pair	0.047	0.090	0.064	0.060

Table 4.14 P values of the impedance profile comparison A versus C for the four cables

A vs C P VALUE	CABLE 1	CCA CABLE 2	CABLE 3	CABLE 4
Orange Pair	0.366	0.901	0.687	0.449
Green Pair	0.520	0.617	0.601	0.692
Blue Pair	0.754	0.690	0.322	0.983
Brown Pair	0.936	0.284	0.674	0.759

The summary of the comparison of the impedance profiles measurements A and C using the KS test for the four cables is illustrated with a chart for the test D values in Figure 4.29 and the P values in 4.30.

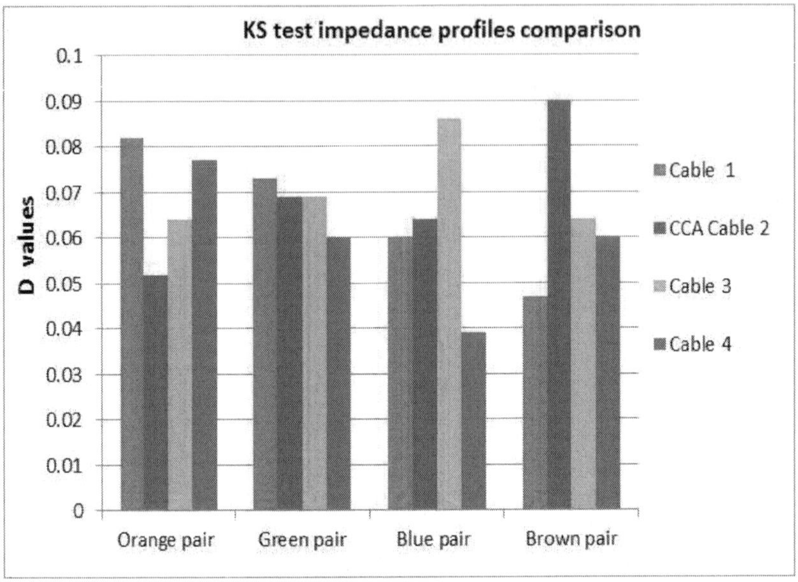

Figure 4.29 KS test D values chart for the impedance profiles of the four cables

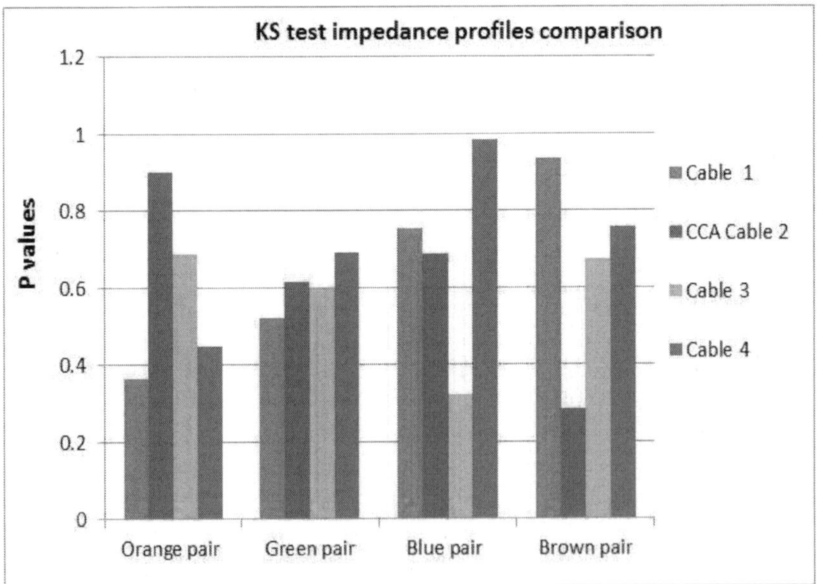

Figure 4.30 KS test P values chart for the impedance profiles of the four cables

The summary of the KS test as shown in Figure 4.29 is that the D values of all the pairs of the four cables are below 0.1260 as presented in Tables 4.13 and 4.14. Similarly, the KS tool results in Figure 4.30 indicates that the P values of all the pairs of the four cables are greater than 0.05 as presented in Tables 4.13 and 4.14. This means that the null hypothesis cannot be rejected or is accepted and therefore the difference between the impedance profiles measurements A and C is not significant.

4.4 Quantifying Variations in NEXT Measurements due to Handling Stress

This section presents a method of evaluating cables performance parameters through their measurements. The method can be used to identify cables with the potential of high susceptibility or otherwise to handling stress expected during installation, particularly reuse operations. The method involves evaluating cables measurements collected through a series of coiling-uncoiling tests to mimic handling stress anticipated during installation or reuse operations. The FSV method and KS test are used to quantify the variations in measurements.

The measurements procedure and terms are presented in section (4.2).

The comparison plots of the NEXT measurements for cables 1 to 4 with Category 6 standard limits [96] are shown in Figures 4.31 to 4.34 using the orange and green pairs combination. The plots in Figures 4.31 to 4.34 show that the NEXT measurements of all the cables do not cross the Category 6 limits.

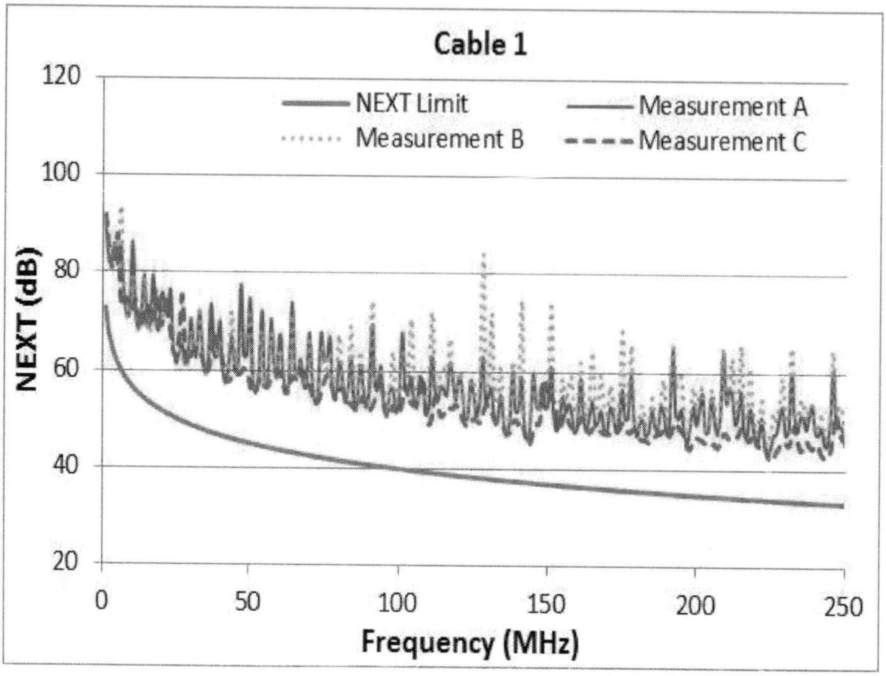

Figure 4.31 Cable 1 NEXT Measurements

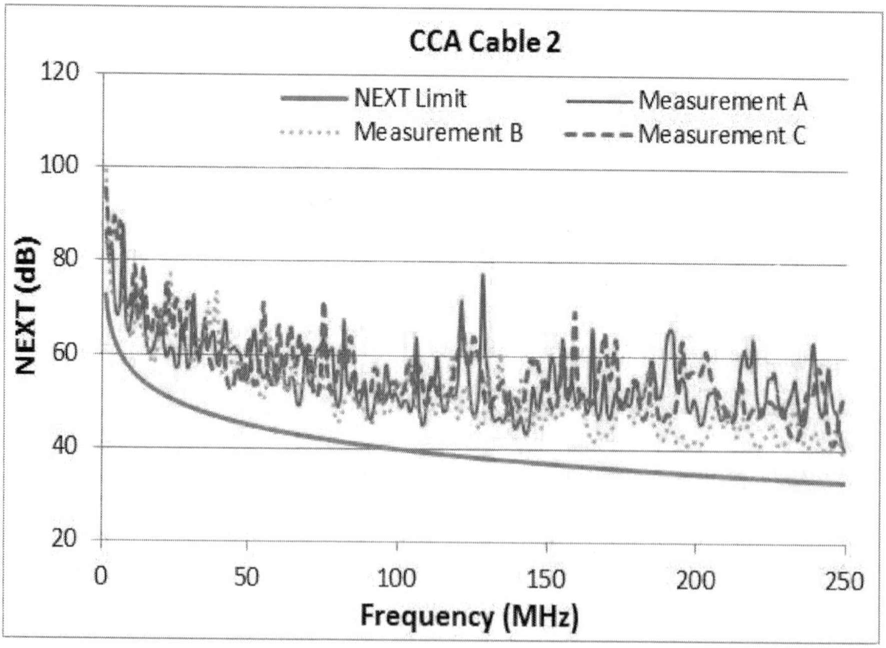

Figure 4.32 CCA Cable 2 NEXT Measurements

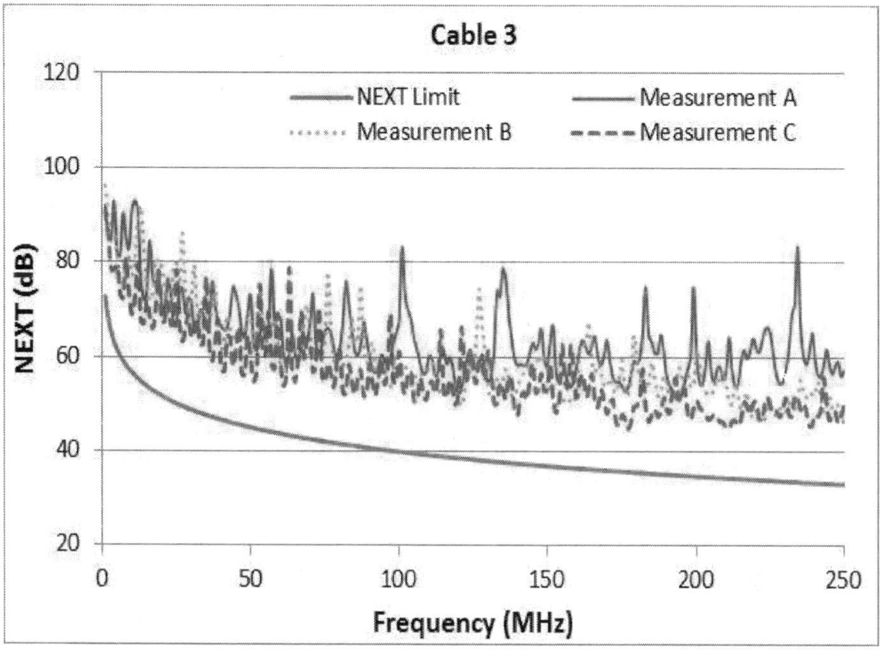

Figure 4.33 Cable 3 NEXT Measurements

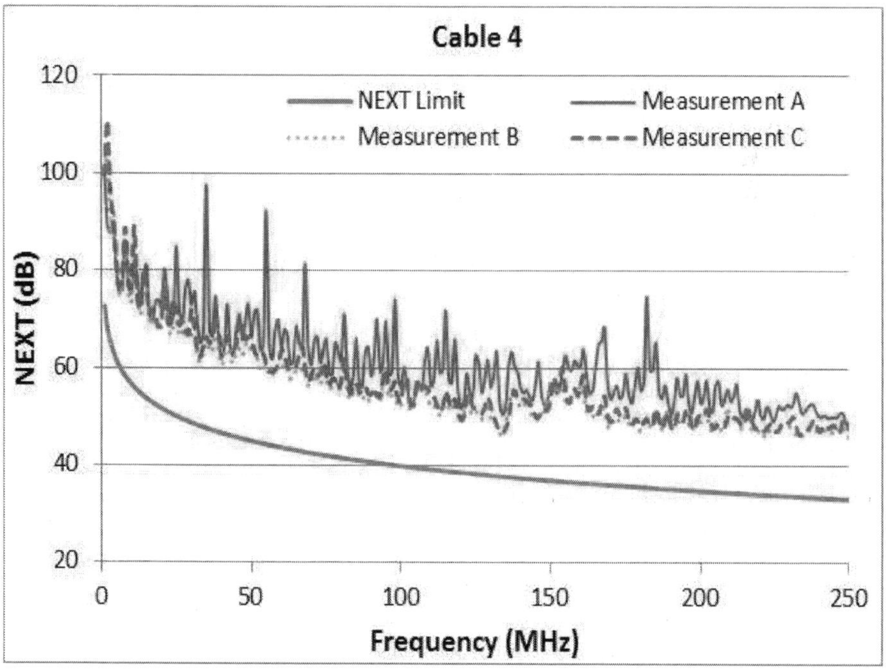

Figure 4.34 Cable 4 NEXT Measurement

4.4.1 Application of FSV to Quantify Variations in NEXT Measurement

This section applies the FSV to quantify variations in NEXT measurements due to the effects of handling as explained in section (3.3). The aim is to quantify the variations in NEXT between the baseline which is the first test (measurement A) and third test (measurement C) for the four Category 6 cables to be evaluated. The results of the analysis will help identify cables that have high resilience to handling stress with the lowest potential of susceptibility to degradation overtime or reuse. The FSV results of the NEXT comparison between measurements A and C of the orange, green, blue and brown pairs for the four cables are presented in Tables 4.15 to 4.17.

Table 4.15 FSV results of the NEXT comparison of the orange and green pairs combination

MEASUREMENT A vs C	ADM_{tot}	FDM_{tot}	GDM_{tot}
CABLE 1	0.2254	0.4395	0.5343
CCA CABLE 2	0.3830	0.4207	0.6294
CABLE 3	0.4021	0.5207	0.7269
CABLE 4	0.3278	0.5273	0.6759

Table 4.16 FSV results of the NEXT comparison of the green and blue pairs combination

MEASUREMENT A vs C	ADM_{tot}	FDM_{tot}	GDM_{tot}
CABLE 1	0.2443	0.3821	0.5009
CCA CABLE 2	0.2860	0.4962	0.6313
CABLE 3	0.2323	0.3797	0.4881
CABLE 4	0.2956	0.4499	0.5937

Table 4.17 FSV results of the NEXT comparison of the blue and brown pairs combination

MEASUREMENT A vs C	ADM_{tot}	FDM_{tot}	GDM_{tot}
CABLE 1	0.3543	0.4374	0.6190
CCA CABLE 2	0.5783	0.7651	1.0537
CABLE 3	0.2795	0.3954	0.5364
CABLE 4	0.3956	0.5161	0.7221

The FSV GDM results in Tables 4.15 to 417 between the baseline (measurement A) and the third test (measurement C) for cable 1 gave the least difference value of 0.5343 for the orange/green pairs combination, while cable 3 gave the least difference values of 0.4881 and 0.5364 for the green/blue and the blue/brown pairs combinations respectively. The FSV results in Tables 4.15 to 4.17 therefore show that cable 1 and cable 3 gave the highest resilience to the stress the cables were subjected to, after the third test. On the other hand, the FSV GDM results in Tables 4.15 to 4.17 indicates that the CCA cable 2 gave the highest variations between the baseline (measurement A) and the third test (measurement C) for the green/blue pairs and the blue/brown pairs combinations with values of 0.6313 and 1.0537 respectively. The FSV results in Tables 4.15 to 4.17 therefore showed that the CCA cable 2 gave the highest susceptibility to the handling stress by providing the highest difference between measurement A and C for two out of the three pairs combinations assessed.

The summary of the FSV results for the comparison between NEXT measurements A and C for the four cables is illustrated with a chart in Figure 4.35.

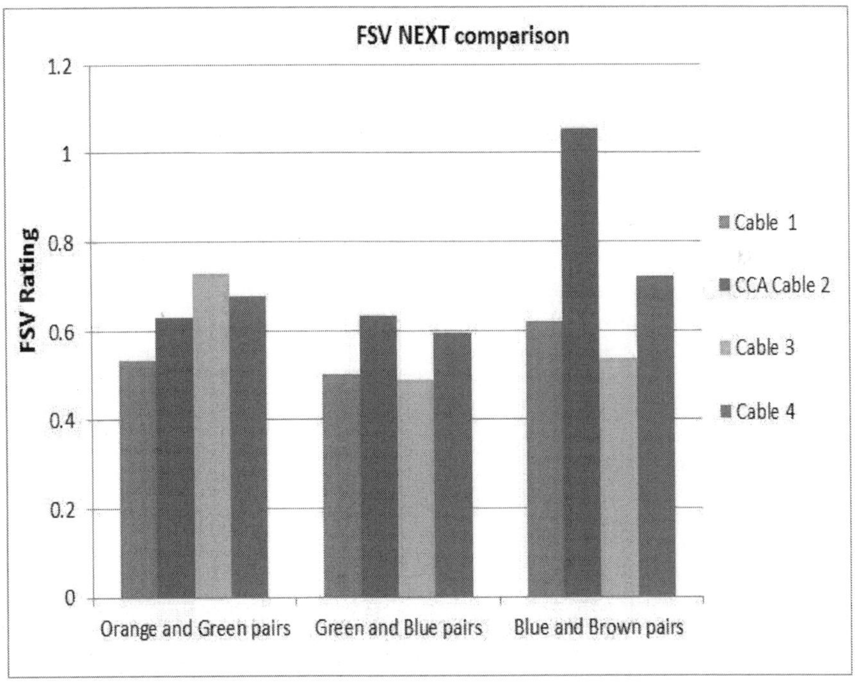

Figure 4.35 FSV NEXT comparison chart for the four cables

The summary of the NEXT analysis using the FSV is that cable 1 (orange/green pairs combination) and cable 3 (green/blue and blue/brown pairs combinations) gave the least variations between the baseline (measurement A) and the third test (measurement C) as shown in Figure 4.35. On the other hand, the CCA cable 2 (green/blue pairs and the blue/brown pairs combinations) gave the highest return loss difference between the baseline (measurement A) and the third test (measurement C) as shown in Figure 4.35. However, the FSV results in Table 4.15 to 4.17 shows that the similarity between the baseline (measurement A) and the third test (measurement C) is fair for all the cables pairs combinations, except the CCA cable that gave a poor difference for the blue/brown pairs combination. This shows that the impact of the whole length coiling and uncoiling actions after the third test. However, since the three rounds of whole cable length coiling and uncoiling actions is a maximal case test, the cables performance is satisfactory, except for the CCA cable 2. The KS tool will be used in the next section to determine whether the impact of the three handing stress tests on the cables is significant or not on the NEXT.

4.4.2 Application of KS Test to Quantify Variations in NEXT Measurement

This subsection applies the KS test to determine whether the handling stress test creates a significant difference or not on the NEXT comparison between measurement A (baseline) versus measurement C (third test). The KS test already discussed in section (2.4.2) uses the P or the test statistic D values to determine the significance or otherwise of the differences between data sets comparison. However, the P value is the main deciding factor as mentioned in section (2.4.2). The critical value (D_{crit}) to be used as baseline for comparison with the test statistic D values was computed using equation (49). In this case, k is 1.36 [77], [82] for a significance value (α) of 0.05, N_1=N_2=250 which gives D_{crit} = 0.1216.

The KS test results for each of the four pairs of cables 1 to 4 is shown in Table 4.7 and Table 4.8 for the test statistic value D and the P value. The conditions for null hypothesis rejection or otherwise is stated below:

If $D > 0.1216$ or $P < 0.05$ means null hypothesis is rejected (significant difference)

If $D < 0.1216$ or $P > 0.05$ means null hypothesis cannot be rejected or is accepted (no significant difference).

The KS test results in Tables 4.18 and 4.19 indicates that cable 1 and cable 4 (green/blue pairs combinations) and cable 3 (blue/brown pairs combinations) showed no significance difference between the baseline (measurement A) and third test (measurement C) comparison as their D values is less than 0.1216 and P values greater than 0.05. However, the CCA cable 2 (all pairs) showed significant difference between measurement A and measurement C comparison as their D values greater than 0.1216 and P values lower than 0.05.

Table 4.18 D values of the NEXT comparison A versus C for the four cables

A vs C D Value	CABLE 1	CCA CABLE 2	CABLE 3	CABLE 4
Orange and Green Pairs	0.2320	0.1960	0.4840	0.2920
Green and Blue Pairs	0.0920	0.1281	0.2120	0.1200
Blue and Brown Pairs	0.2160	0.7800	0.0840	0.1640

Table 4.19 P values of the NEXT comparison A versus C for the four cables

A vs C P Value	CABLE 1	CCA CABLE 2	CABLE 3	CABLE 4
Orange and Green Pairs	0.000	0.000	0.000	0.000
Green and Blue Pairs	0.229	0.000	0.000	0.053
Blue and Brown Pairs	0.000	0.000	0.327	0.002

The summary of the comparison of the return loss measurements A and C using the KS test for the four cables is illustrated with a chart for the test D values in Figure 4.36 and the P values in 4.37.

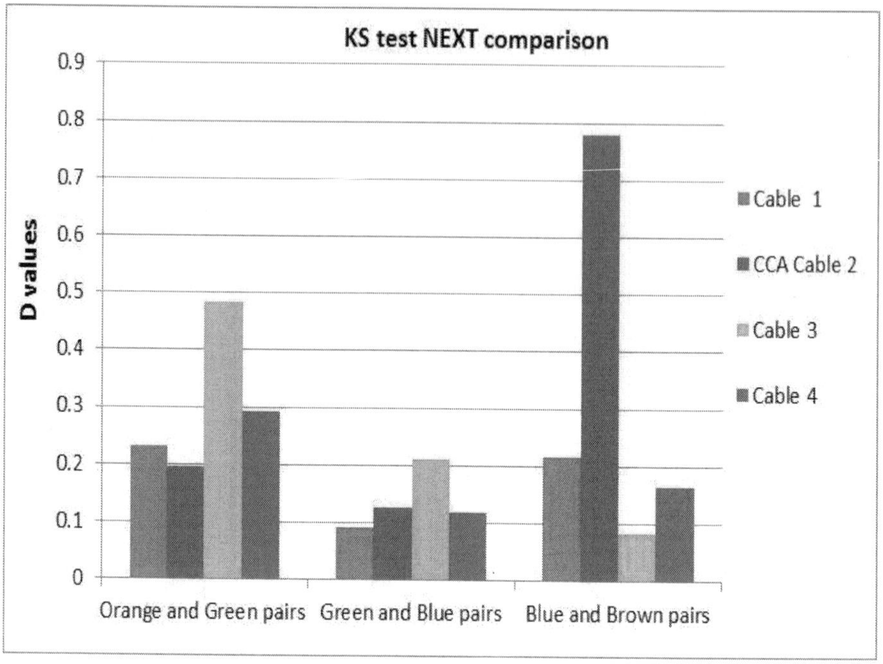

Figure 4.36 KS test D values chart for the NEXT of the four cables

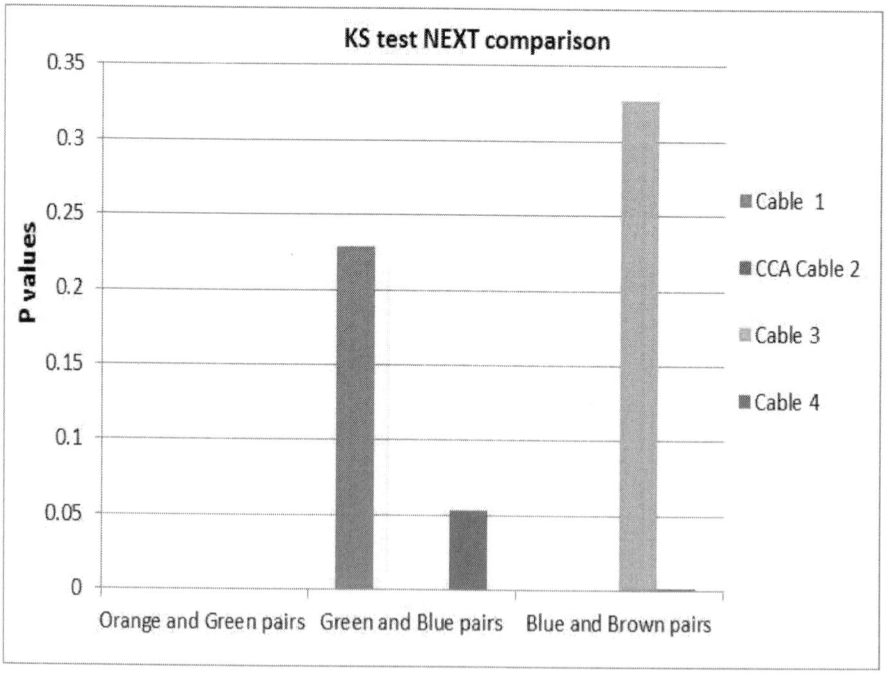

Figure 4.37 KS test P values chart for the NEXT of the four cables

The summary of the KS test results for NEXT presented in Figures 4.36 and 4.37 is that cable 1, cable 3 and cable 4 gave the best result between measurement A (first test) and measurement C (third test) comparison as they showed no significant difference in one out of the three pairs combinations tested as their D values are less than 0.1216 and P values greater than 0.05. On the other hand, the CCA cable 2 provided the worst results as it gave a significant difference between measurement A and measurement C comparison in all the three pairs combinations tested as their D values are greater than 0.1216 and P values are below 0.05 as shown in Figures 4.36 and 4.37.

4.5 Quantifying Variations in RLGC Parameters due to Handling Stress

In this section, the effects of handling stress on the cable structure (RLGC) parameters of the four Category 6 UTP cables from different manufacturers already mentioned in previous sections was examined. The RLGC parameters of the four Category 6 cables were obtained from their impedance profile measurements using the method explained in section (2.1.6) and equations (28) to (38). To evaluate the RLGC parameters due to handling stress, the cables impedance measurements taken as explained in sections (3.1) and (3.3) was used as the cable impedance (Zo) in equations (35) to (38). The measurements procedure and terms for the impedance profiles are hereby presented again as follows:

Measurement A: UTP cables used to form coils of about 30cm diameter and then stretched out before measurement

Measurement B: UTP cables used for measurements A, reused to form coils of about 30cm diameter and then stretched out before measurement

Measurement C: UTP cables used for measurements B, reused to form coils of about 30cm diameter and then stretched out before measurement

The cables dimensions are:

Cable 1: D=0.99 mm, d=0.57 mm

CCA Cable 2: D=1.03 mm, d=0.57 mm

Cable 3: D=0.96 mm, d=0.54 mm

Cable 4: D=1.01 mm, d=0.57 mm

D is the Distance between the centers of the two conductors and d is the diameter of the conductors.

The cable material properties are:

Relative permittivity of dielectric material (ε_r = 2.3) [97], [98] for polyethylene,

Permittivity of free space (ε_o) = 8.8542 x 10–12 F/m,

Conductivity of conductor (σ_c) = 5.8 x 107S/m for copper [39],

Permeability of conductor:$\mu_c = \mu_o = 4\pi$ x 10–7H/m [39],

Permittivity of dielectric:$\varepsilon = \varepsilon_o \times \varepsilon_r$ [39].

The resistance per unit length using the orange pair measurements for cables 1 to 4 computed with equations (26) to (38) is presented in Figures 4.38 to 4.41. A view of the plots in Figures 4.38 to 4.41 shows that it is the only the CCA cable 2 that crosses the 4.2Ω/m point, others are below the 4.0Ω/m point.

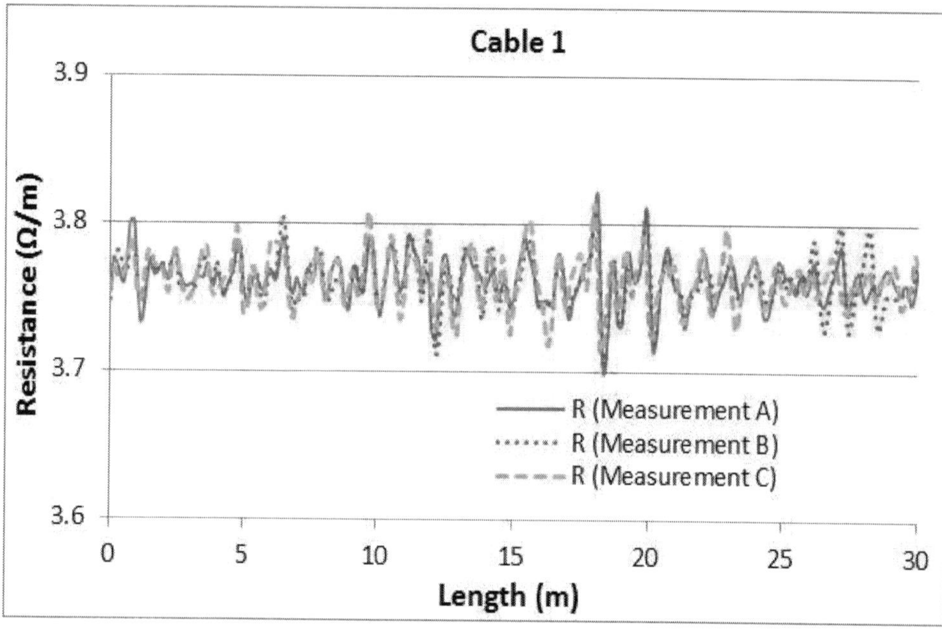

Figure 4.38 Resistance plot for cable 1

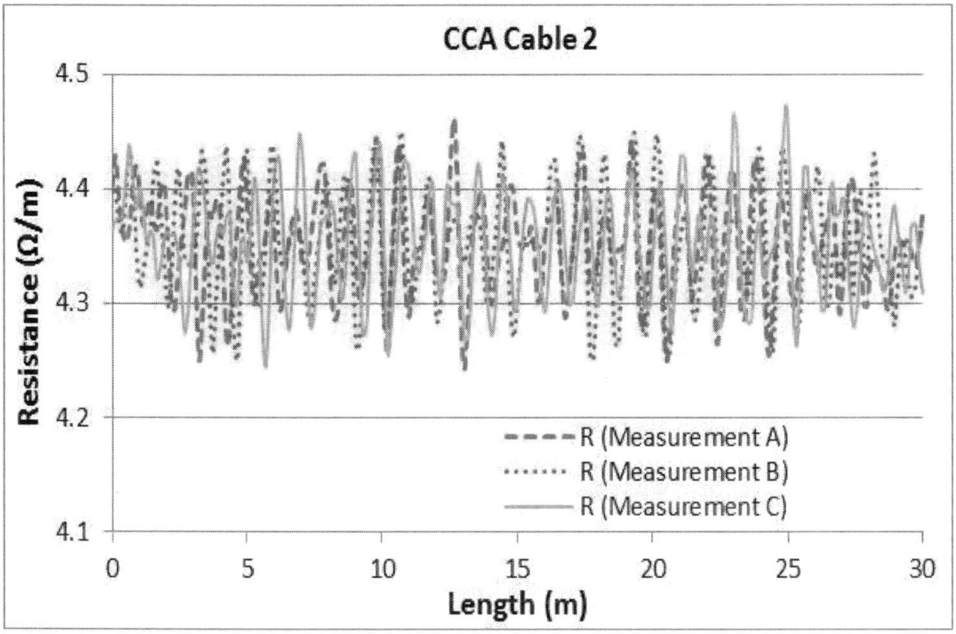

Figure 4.39 Resistance plot for CCA cable 2

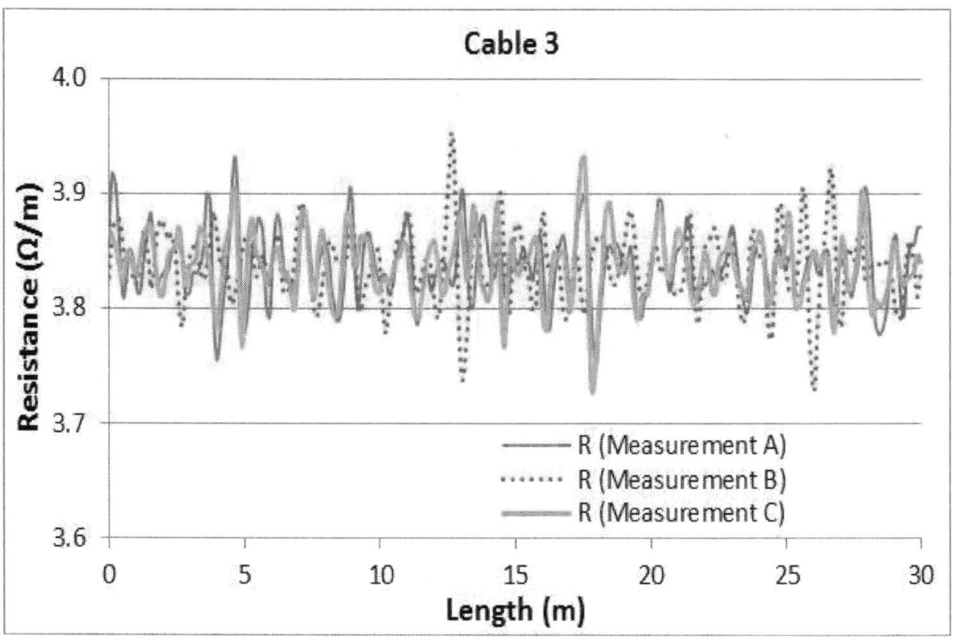

Figure 4.40 Resistance plot for cable 3

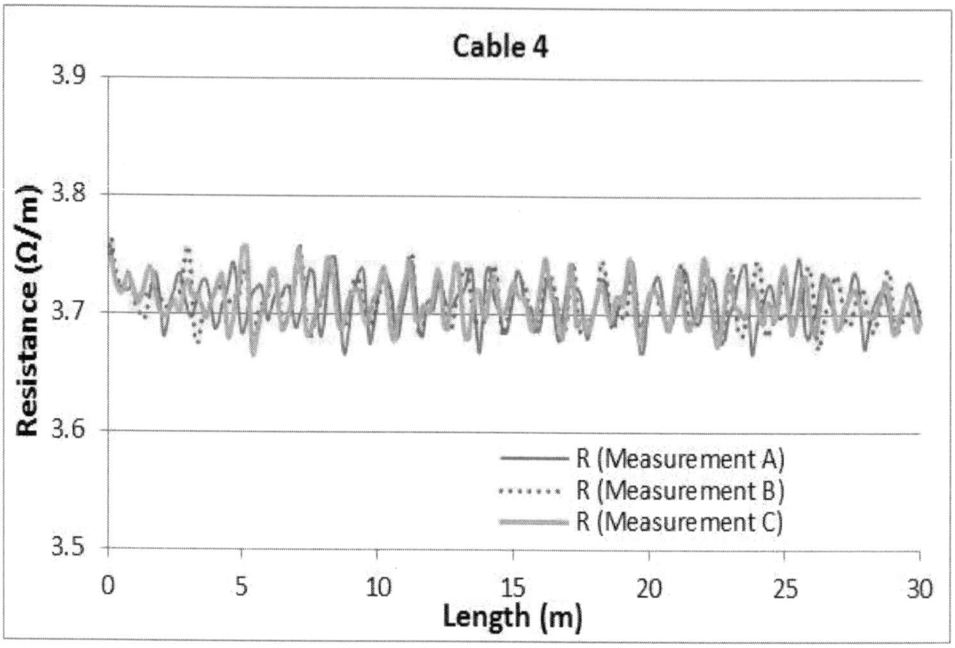

Figure 4.41 Resistance plot for cable 4

4.5.1 Application of FSV to Quantify Variations in RLGC Parameters

A view of Figures 4.38 to 4.41 shows that it is very difficult to evaluate the similarity between the RLGC parameters using the human eye. This difficulty indicates the need for the use of the FSV to evaluate the RLGC parameters from the first and third handling impedance profile measurement tests. An evaluation of the RLGC parameters of the first test (measurement A) in comparison with the third test (measurement C) using the FSV was therefore carried out to determine the effects of handling stress on the cable structures. The FSV results of the comparison between the baseline (first test) and third test of the RLGC parameters of the orange and green pairs for the four cables examined are shown in Tables 4.20 and 4.21.

An analysis of the FSV results in Tables 4.20 and 4.21 of the RLGC parameters for the first test (measurement A) and third test (measurement C) comparison of the orange and green pairs indicates that they follow the pattern of the FSV comparison results of the impedance profile measurements A versus C presented in Tables 4.9 and 4.10. The RLGC computation for the blue and brown pairs was therefore discontinued. The FSV results in Tables 4.20 and 4.21 indicate that the cable impedance has a great influence on the RLGC parameters by dictating their outputs. The FSV results in Tables 4.20 and 4.21 shows that cable 1 gave the least variations between measurements A and C comparison, while the CCA cable 2 (orange pair) and cable 4 (green pair) gave the highest variations between measurements A and C comparison.

Table 4.20 FSV RLGC comparison results for the orange pair of the four cables

RESISTANCE (R) A vs C	ADM_{tot}	FDM_{tot}	GDM_{tot}
CABLE 1	0.2977	0.3353	0.5101
CCA CABLE 2	0.3493	0.4052	0.5886
CABLE 3	0.2825	0.3314	0.4872
CABLE 4	0.3227	0.3480	0.5287
INDUCTANCE (L) A vs C	ADM_{tot}	FDM_{tot}	GDM_{tot}
CABLE 1	0.2977	0.3353	0.5101
CCA CABLE 2	0.3493	0.4052	0.5886
CABLE 3	0.2825	0.3314	0.4872
CABLE 4	0.3227	0.3480	0.5287
CONDUCTANCE (G) A vs C	ADM_{tot}	FDM_{tot}	GDM_{tot}
CABLE 1	0.2979	0.3354	0.5101
CCA CABLE 2	0.3496	0.4045	0.5884
CABLE 3	0.2815	0.3301	0.4854
CABLE 4	0.3226	0.3481	0.5288
CAPACITANCE (C) A vs C	ADM_{tot}	FDM_{tot}	GDM_{tot}
CABLE 1	0.2979	0.3354	0.5101
CCA CABLE 2	0.3496	0.4045	0.5884
CABLE 3	0.2815	0.3301	0.4854
CABLE 4	0.3226	0.3481	0.5288

Table 4.21 FSV RLGC comparison results for the green pair of the four cables

RESISTANCE (R) A vs C	ADM_{tot}	FDM_{tot}	GDM_{tot}
CABLE 1	0.3103	0.3745	0.5479
CCA CABLE 2	0.3521	0.3790	0.5692
CABLE 3	0.3413	0.3567	0.5490
CABLE 4	0.3897	0.4432	0.6570

INDUCTANCE (L) A vs C	ADM_{tot}	FDM_{tot}	GDM_{tot}
CABLE 1	0.3103	0.3745	0.5479
CCA CABLE 2	0.3521	0.3790	0.5692
CABLE 3	0.3413	0.3567	0.5490
CABLE 4	0.3897	0.4432	0.6570
CONDUCTANCE (G) A vs C	ADM_{tot}	FDM_{tot}	GDM_{tot}
CABLE 1	0.3112	0.3678	0.5402
CCA CABLE 2	0.3522	0.3793	0.5694
CABLE 3	0.3415	0.3561	0.5488
CABLE 4	0.3895	0.4430	0.6568
CAPACITANCE (C) A vs C	ADM_{tot}	FDM_{tot}	GDM_{tot}
CABLE 1	0.3112	0.3678	0.5403
CCA CABLE 2	0.3522	0.3793	0.5694
CABLE 3	0.3415	0.3561	0.5488
CABLE 4	0.3896	0.4431	0.6568

The summary of the FSV RLGC comparison results in Tables 4.20 and 4.21 indicates that it follows the same output pattern of the FSV impedance profile results in Tables 4.9 and 4.10. It also indicate that cable 1 (orange pair) and cable 3 (green pair) gave the highest resilience to the series of handling stress tests. On the other hand, the CCA cable 2 (orange pair) and cable 4 (green pair) gave the lowest resilience the series of handling stress tests.

4.5.2 Application of the KS Test to Quantify Variations in RLGC Parameters

This section applies the KS tool to determine whether the impact of the handling stress tests on the cables is significant or not on the RLGC parameters (measurement A versus measurement C) comparison. The critical value (D_{crit}) was computed using equation (49). In this case, N_1 and N_2 are the length of the data sets to be compared; k is 1.36 [77], [82] for a significance value (α) of 0.05 and $N_1 = N_2 = 233$ which gives $D_{crit} = 0.126$. As stated in section (2.4.2), the null hypothesis cannot be rejected or is accepted if the P value from the KS test is greater than the significance (α) or the test statistic value D is less than the critical value computed for the data sets and vice versa. This can be explained as follows:

If $D > 0.126$ or $P < 0.05$ means null hypothesis is rejected (significant difference)

If D < 0.126 or P > 0.05 means null hypothesis cannot be rejected or is accepted (no significant difference). The KS test results for RLGC comparison between measurements A and C for the orange pair is shown in Tables 4.22 and 4.23 for the test statistic values D and P values, while that for the green pair is shown in Table 4.24 and 4.25. The KS test RLGC results in Tables 4.22 to 4.25 for the orange and green pairs shows that they follow the output pattern of the KS test results of the impedance profile for the green and orange pairs in Tables 4.13 and 4.14. This indicates the influence of the cables impedances due to handling stress tests used in the computation of the RLGC parameters. The results of the KS test in Tables 4.22 and 4.24 shows that the D values of the four cables are below 0.126. Similarly, the KS test results in Table 4.23 and 4.25 indicates that the P values of all the four pairs of the four cables are greater than 0.05 meaning that the null hypothesis cannot be rejected or is accepted. The results indicates that the difference between the RLGC parameters for measurements A and C comparison is not significant.

Table 4.22 D values of the RLGC comparison of the orange pair for the four cables

D VALUE (A vs C)	CABLE 1	CCA CABLE 2	CABLE 3	CABLE 4
RESISTANCE (R)	0.082	0.052	0.064	0.077
INDUCTANCE (L)	0.082	0.052	0.064	0.077
CONDUCTANCE (G)	0.082	0.052	0.064	0.077
CAPACITANCE (C)	0.082	0.052	0.064	0.077

Table 4.23 P values of the RLGC comparison of the orange pair for the four cables

P VALUE (A vs C)	CABLE 1	CCA CABLE 2	CABLE 3	CABLE 4
RESISTANCE (R)	0.366	0.901	0.687	0.449
INDUCTANCE (L)	0.366	0.901	0.687	0.449
CONDUCTANCE (G)	0.366	0.901	0.687	0.449
CAPACITANCE (C)	0.406	0.910	0.706	0.475

Table 4.24 D values of the RLGC comparison of the green pair for the four cables

D VALUE (A vs C)	CABLE 1	CCA CABLE 2	CABLE 3	CABLE 4
RESISTANCE (R)	0.073	0.069	0.069	0.060
INDUCTANCE (L)	0.073	0.069	0.069	0.060
CONDUCTANCE (G)	0.073	0.069	0.069	0.060
CAPACITANCE (C)	0.073	0.069	0.069	0.060

Table 4.25 P values of the RLGC comparison of the green pair for the four cables

P VALUE (A vs C)	CABLE 1	CCA CABLE 2	CABLE 3	CABLE 4
RESISTANCE (R)	0.520	0.617	0.601	0.692
INDUCTANCE (L)	0.520	0.617	0.601	0.692
CONDUCTANCE (G)	0.520	0.617	0.601	0.692
CAPACITANCE (C)	0.549	0.627	0.612	0.782

The summary of the KS test results in Tables 4.22 to 4.25 is that the difference between measurements A and C comparison for the RLGC parameters is not significant as the P values of all the cables are greater than 0.05 and D values are less than 0.126. The KS test result shows that the effect of the series of handling stress tests on the cables is not significant on their RLGC parameters.

4.6 Summary of the Cables Measurements Assessment due to Handling Stress

This section gives a summary of the measurements assessment results obtained due to handling stress in section 4.2, 4.3 and 4.4 for return loss, impedance profile and NEXT respectively. These are outlined as follows:

Return Loss Measurement

The return loss measurement assessment due to handling stress shows that cable 1 gave the best result between measurement A (first test) and measurement C (third test) comparison as it showed no significant difference in three of it pairs (orange, blue and brown) as their P values are greater than 0.05 and D values are less than 0.067 (baseline). On the other hand, the CCA cable 2 provided the worst results as it gave a significant difference between measurement A and measurement C comparison for all the pairs as their P values are below 0.05 and D values greater than 0.067 (baseline).

Impedance Profile

The summary of the impedance profiles comparison between measurement A (first test) and measurement C (third test) is that the D values of all the pairs of the four cables are greater than 0.05 and D values are below 0.126 (baseline). This means that the null hypothesis cannot be rejected or is accepted and therefore the difference between the impedance profiles measurements A and C is not significant for all the pairs of the four cables.

Near-end Crosstalk

The summary of the KS test results for NEXT is that cable 1, cable 3 and cable 4 gave the best result between measurement A (first test) and measurement C (third test) comparison as they showed no significant difference in one out of the three pairs combinations tested as their D values are less than 0.1216 and P values greater than 0.05. On the other hand, the CCA cable 2 provided the worst results as it gave a significant difference between measurement A and measurement C comparison in all the three pairs combinations tested as their P values are less than 0.05 and D values are greater than 0.1216 (baseline).

Chapter 5

CROSSTALK EVALUATION AND MODELING

This chapter evaluates the NEXT predictions of the standard simple and advanced models explained in section (2.5) using the ANSI and average crosstalk parameters (constants) of three Category 6 UTP cables from different manufacturers. The ANSI and average crosstalk constants are the limit line determinants in the models and will be used as starting points in the investigation of better crosstalk parameters in UTP cables using Category 6 cables as an example. Crosstalk is caused by unwanted signal coupling between pairs in the same channel or cable bundle [8].

5.1 Near End Crosstalk Prediction using the Standard Simple Model

This section evaluates NEXT using the standard simple model with the ANSI and average crosstalk constant obtained from the three Category 6 cables under examination. The aim is to provide an improved crosstalk constant that can be used to predict NEXT in Category 6 cables using the standard simple model. The crosstalk constant of each cable pair was obtained using equation (60) in section (2.5.4) which is again presented here as:

$$K_{NEXT} = 10^{-\left(\frac{A_{NEXT}(f) + 15\log f}{10}\right)} \tag{62}$$

where, $A_{NEXT}(f)$ is the measured crosstalk attenuation and f is the frequency in Hz.

The simplified NEXT formula already given in section (2.5.2) is again presented [23], [89] as:

$$NEXT(dB) = -10\log_{10}(K_{NEXT}f^{3/2}) \tag{63}$$

The $f^{3/2}$ parameter was formulated in [23] by ANSI and FSAN after a series of twisted pair cable measurements.

The average crosstalk constant of each cable is presented in Table 5.1.

Table 5.1 Average crosstalk constants of the cables

CABLE	K_{NEXT}
1	2.6755 x 10–16
2	1.5481 x 10–16
3	2.6872 x 10–16
AVERAGE	2.3036 x 10–16

The plots of the comparison of the simulation of the standard simple NEXT using the ANSI and the average crosstalk constant limits of the cables as shown in Table 5.1 with measurements of the three cables are shown in Figures 5.1 to 5.3.

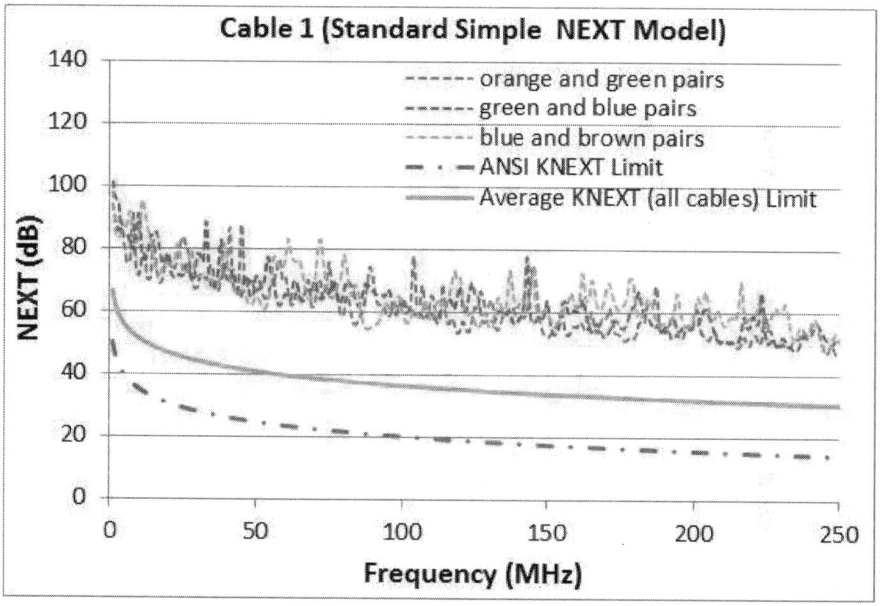

Figure 5.1 Cable 1 NEXT measurements comparison with simulation of the standard simple model using the ANSI and average crosstalk constants

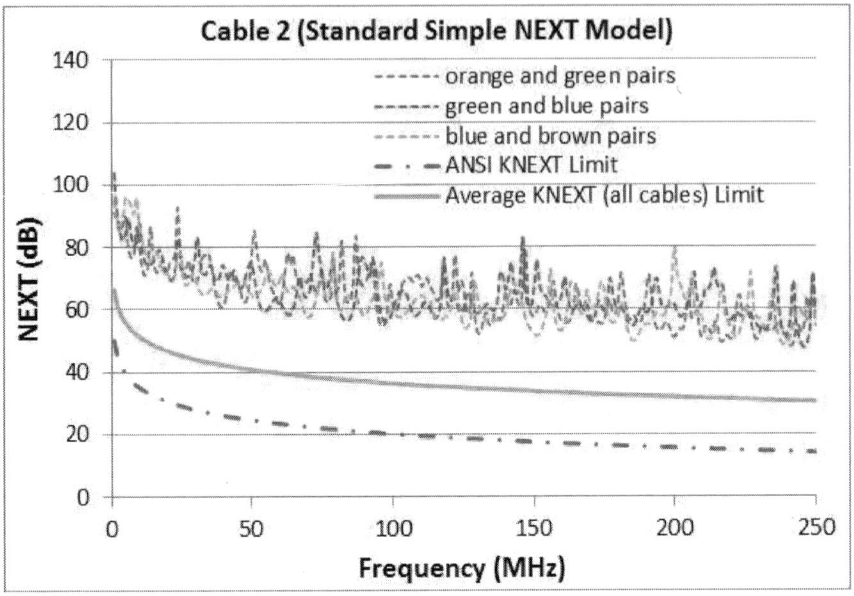

Figure 5.2 Cable 2 NEXT measurements comparison with simulation of the standard simple model using the ANSI and average crosstalk constants

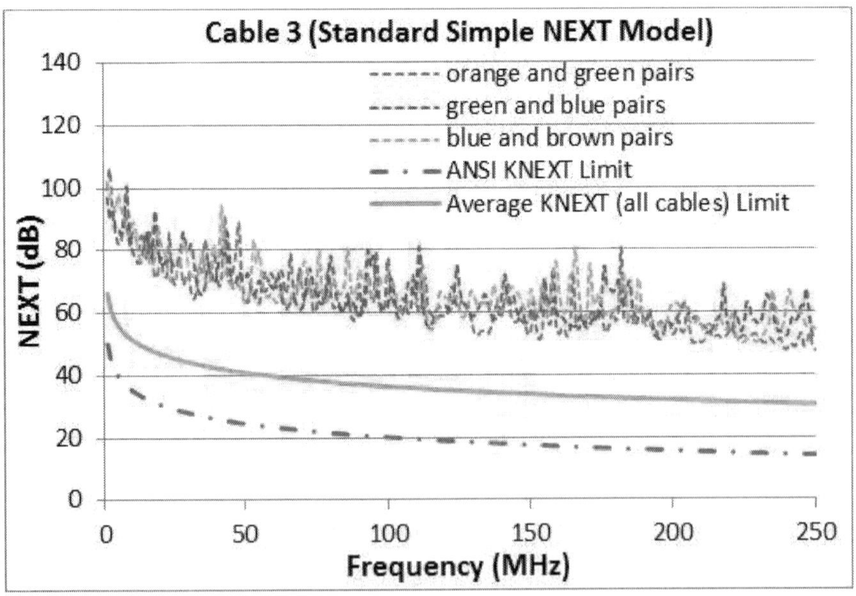

Figure 5.3 Cable 3 NEXT measurements comparison with simulation of the standard simple model using the ANSI and average crosstalk constants

A view of Figures 5.1 to 5.3 shows that using the ANSI and average crosstalk limits of the cables with the standard simple model under predicts the NEXT measurements of the three cables. This indicates that there is need for an improved average crosstalk limit for use with the standard simple model in the NEXT prediction of the Category 6 cables.

The method of curve fitting using the trendline [99] in MS Excel will be used to determine the equation of fit for the NEXT data in combination with the method of simplifying formulas for curve fitting given in [100], [101] to find a suitable K_{NEXT} value for each pair measurements. The logarithm trendline was selected due to the nature of the data and the equation to be used in predicting it. The NEXT formula in equation (63) above can be converted into the form y (NEXT measurements) and x (frequency) with K_{NEXT} represented as K:

$$y = -10log_{10}\left(Kx^{3/2}\right) \quad (64)$$

$$y = -10\log_{10}K - 10\log_{10}x^{3/2} \quad (65)$$

$$y = -10log_{10}K - 10log_{10}x \quad (66)$$

Equation (66) needs to be expressed in the form:

$$Y = Bx + A \text{ [100], [101]} \quad (67)$$

$$y = -15\log_{10}x - 10\log_{10}K \quad (68)$$

$$A = -10\log_{10}K \quad (69)$$

For example, the orange and green pair's measured data of cable 1 gave a trendline equation of the form:

$$y = -8.561\text{In}(x) + 218.71 \quad (70)$$

From equation (70), A = 218.71, this implies that:

$$218.71 = -10\log_{10}K \quad (71)$$

Therefore, K= 1.345860×10^{-22} from equation (71).

The process was repeated for each of the three pairs of cables 1 to 3. The average K_{NEXT} value of the three cables was computed to be (6.589889×10^{-19}) and was taken to be the proposed crosstalk constant.

The plots of the simulation of the standard simple NEXT model using a crosstalk limit of (6.589889×10^{-19}) are shown in Figures 5.4 to 5.6. The plot in Figures 5.4 to 5.6 indicates that the proposed crosstalk constant of (6.589889×10^{-19}) gave an improved prediction of NEXT across the cables.

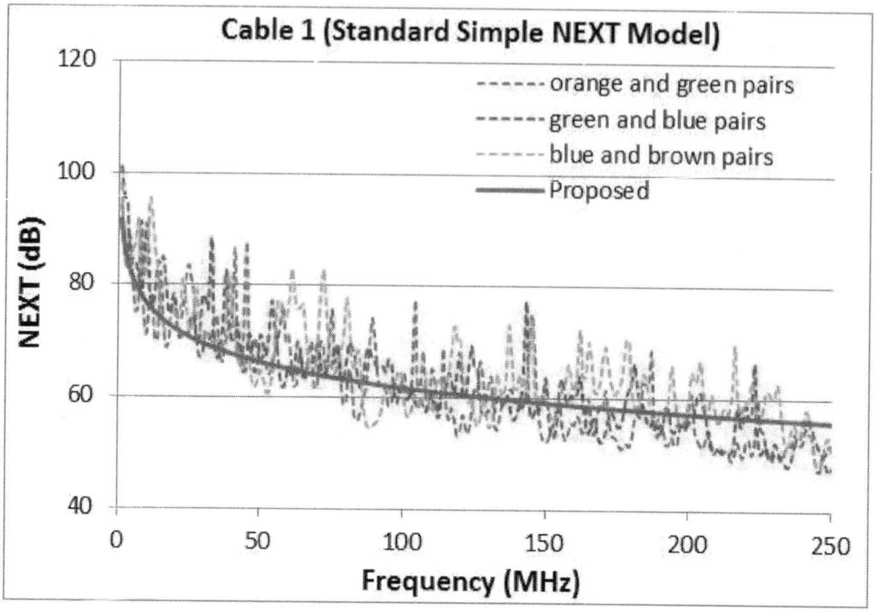

Figure 5.4 Cable 1 NEXT measurements comparison with simulation of the standard simple model using the proposed crosstalk constant

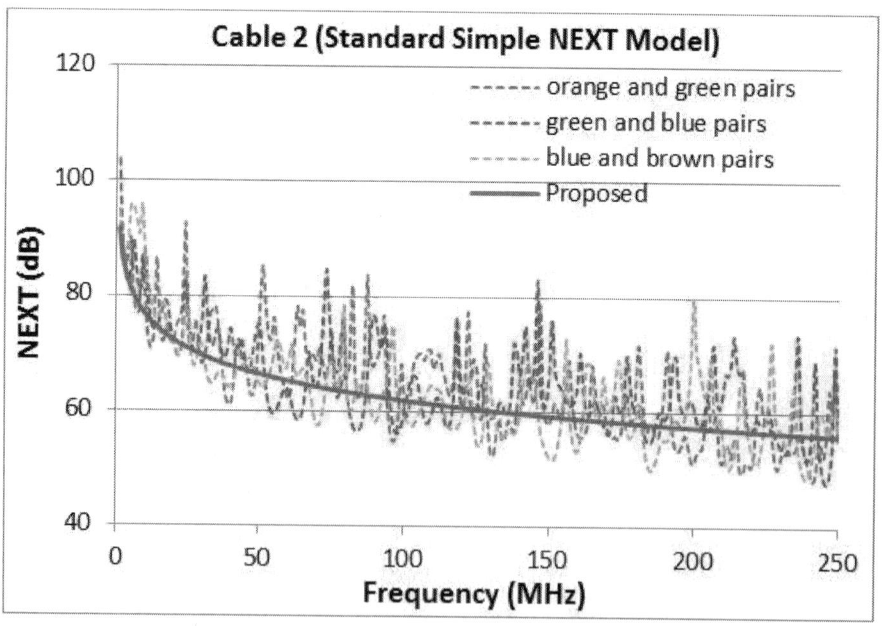

Figure 5.5 Cable 2 NEXT measurements comparison with simulation of the standard simple model using the proposed crosstalk constant

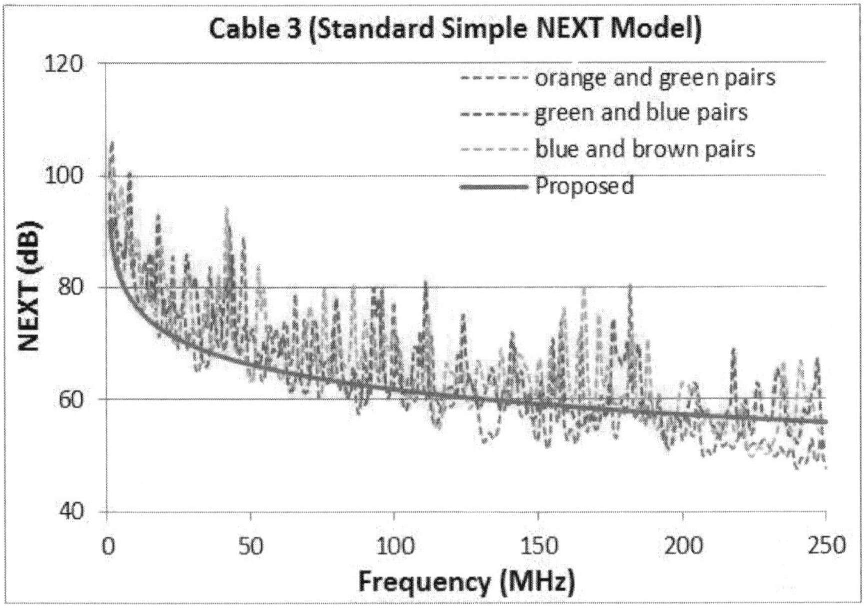

Figure 5.6 Cable 3 NEXT measurements comparison with simulation of the standard simple model using the proposed crosstalk constant

The graphs in Figures 5.1 to 5.6 validates the statement in literature [21], [22] that the standard simple models are not too accurate as they could not predict the peaks and dips often found in NEXT measurements which led to the formulation of the advanced crosstalk model. The next section will therefore evaluate the advanced NEXT model.

5.2 Near End Crosstalk Prediction using the Advanced Model

The advanced NEXT model used in this research was formulated by adapting the approach for the advanced FEXT model presented in [22] and described in section (2.5.4). The advanced NEXT model was presented in section (2.5.4) as:

$$|H_{NEXT}(f)|^2 = K_{NEXT} f^{3/2}\, K_{NORM} \left(\cos\frac{2\pi}{5\lambda} - \cos\frac{9\pi}{10\lambda}\right) \tag{72}$$

The advanced NEXT in decibels can be expressed as:

$$NEXT(dB) = -10\log_{10}|H_{NEXT}(f)|^2 = -10\log_{10}[K_{NEXT} f^{3/2} K_{NORM}\left(\cos\frac{2\pi}{5\lambda} - \cos\frac{9\pi}{10\lambda}\right)] \tag{73}$$

The wavelength is given as $\lambda = \frac{c}{f.\sqrt{\varepsilon_r}}$ (m) (74)

where, f is the frequency in Hz, c is the velocity of light in vacuum and ε_r is the relative permittivity of the cables insulation material.

$K_{NORM} = 3/40$ for UTP cables [22], c = 299792458 m/s and $\varepsilon_r = 2.3$ [97].

The plots of the simulation of the advanced NEXT model given in equation (73) using the average crosstalk constant of all the cables is shown in Figures 5.7 to 5.9.

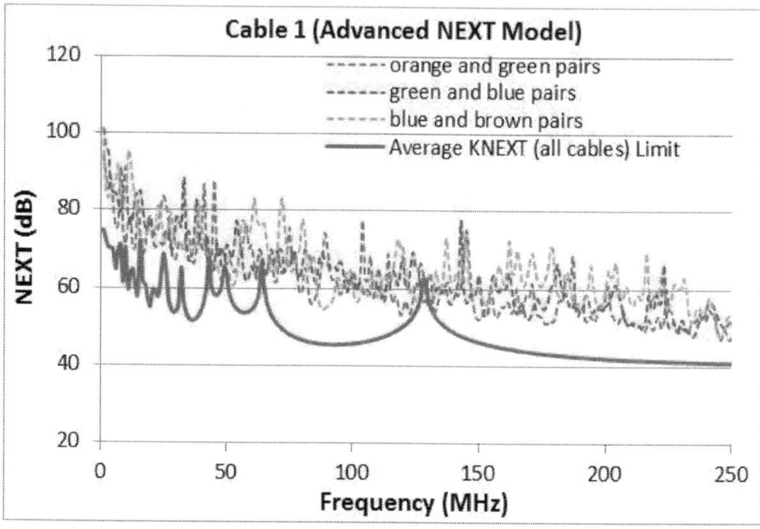

Figure 5.7 Cable 1 NEXT measurements comparison with simulation of the advanced model using the average crosstalk constant

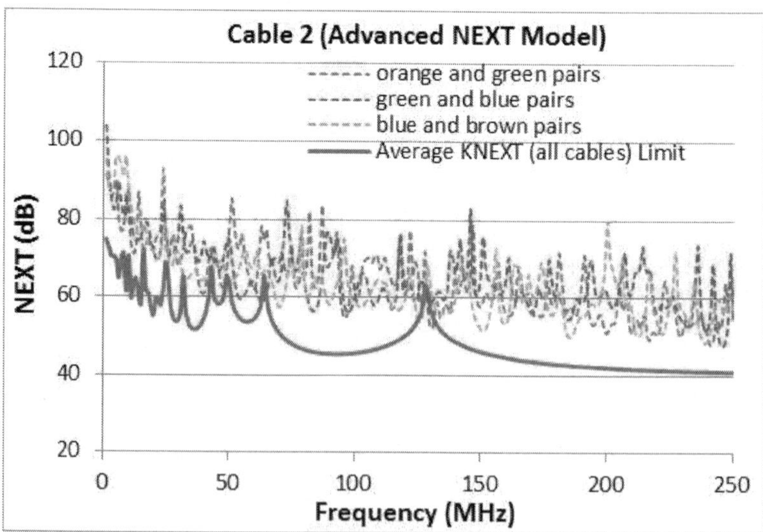

Figure 5.8 Cable 2 NEXT measurements comparison with simulation of the advanced model using the average crosstalk constant

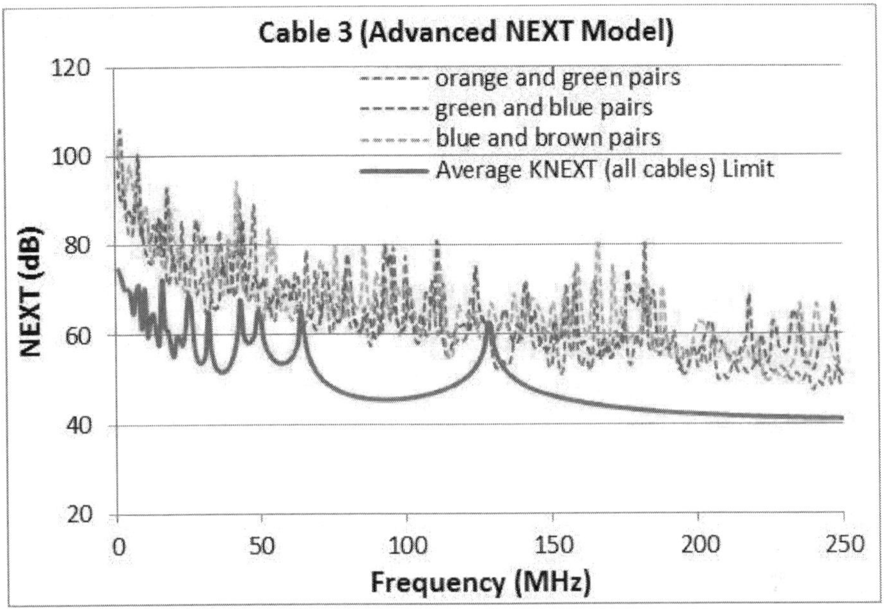

Figure 5.9 Cable 3 NEXT measurements comparison with simulation of the advanced model using the average crosstalk constant

An observation of the plots in Figures 5.7 to 5.9 shows that the simulation of the advanced NEXT model in equation (73) using the average crosstalk constant (all cables) gave a smooth curve that did not predict the peaks and dips of the measurements. This indicates that there is the need for a modification in the cosine function of the NEXT prediction model in equation (73) above and an improved K_{NEXT} constant. The curve fit method using the MS Excel logarithm trendline and the equation of the advanced NEXT model will be used to determine the improved K_{NEXT} and modification (M) required. The advanced NEXT formula with the K_{NEXT} (K) and modification (M) can be expressed as:

$$\text{NEXT(dB)} = -10\log_{10}[\text{K} \times f^{3/2} \times \frac{3}{40} \times \left(\cos\frac{2\pi}{5\lambda M} - \cos\frac{9\pi}{10\lambda M}\right)] \quad (75)$$

Equation (75) can be expressed in the form of y and x to enable logarithm trendline curve fit as:

$$y = -10\log_{10}[\text{K}\, x^{3/2}\, \frac{3}{40} \left(\cos\frac{2\pi}{5\lambda M} - \cos\frac{9\pi}{10\lambda M}\right)] \quad (76)$$

Simplifying the expression in equation (76) with the aim of obtaining the form Y = Bx+ A given in [100], [101] for curve fitting data:

$$y = -10\log_{10}\left(\frac{3}{40}\text{K}x^{3/2}\right) - 10\log_{10}\left(\cos\frac{2\pi}{5\lambda M} - \cos\frac{9\pi}{10\lambda M}\right) \quad (77)$$

$$y = -10\log_{10}\left(\frac{3}{40}K\right) - 10\log_{10}x^{3/2} - 10\log_{10}\left(\cos\frac{2\pi}{5\lambda M} - \cos\frac{9\pi}{10\lambda M}\right) \quad (78)$$

To find K: $y = -10\log_{10}\left(\frac{3}{40}K\right)$ (79)

To find M: $y = -10\log_{10}\left(\cos\frac{2\pi}{5\lambda M} - \cos\frac{9\pi}{10\lambda M}\right)$ (80)

Using cable 1 orange and green pair's measured data as an example, the logarithm trendline equation in MS Excel [99] gave the equation of curve fit as:

$$y = -8.561\text{In}(x) + 218.71 \quad (81)$$

Linking equation (79) with equation (81) using the form:Y = Bx + A, A =218.71.

$$218.71 = -10\log_{10}\left(\frac{3}{40}K\right) \quad (82)$$

$$K=1.794480 \times 10^{-21} \quad (83)$$

Similarly, solving M in equation (80) with f=1MHz and λ as given in equation (74):

$$218.71 = -10\log_{10}\left(\cos\frac{2\pi}{5\lambda M} - \cos\frac{9\pi}{10\lambda M}\right) \quad (84)$$

$$21.871 = -\log_{10}\left(\cos\frac{6.2832}{988.39M} - \cos\frac{28.27}{1976.77M}\right) \quad (85)$$

$$21.871 = -\log_{10}\left(\cos\frac{0.0064}{M} - \cos\frac{0.0143}{M}\right) \quad (86)$$

$$\text{antilog}(-21.871) = \left(\cos\frac{0.0064}{M} - \cos\frac{0.0143}{M}\right) \quad (87)$$

$$M=1.850641 \times 10^{14} \quad (88)$$

The process was repeated for other pairs considered and the average K and M was determined as in equations (89) and (90).

The average $K=1.620455 \times 10^{-17}$ (89)

The average $M=7.437277 \times 10^{14}$ (90)

The plots of the introduction of M with no M in the cosine function using the average crosstalk (K_{NEXT}) limit for all the cables is shown in Figures 5.10 to 5.12 for cables 1 to 3. The plots in Figures 5.10 to 5.12 indicates that the use of no M gave only a smooth curve across the measurements unlike the use of M that gave a curvy and wavy pattern typical of most NEXT measurements. However, this needs the use of the improved K_{NEXT} to predict the NEXT measurements.

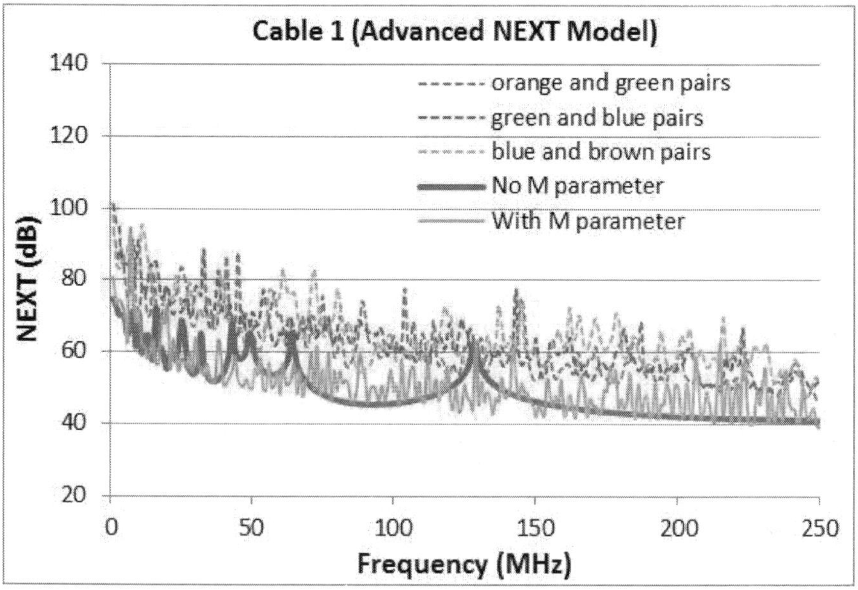

Figure 5.10 Cable 1 NEXT measurements comparison with simulation of the advanced model using the average crosstalk constant with M and no M parameters

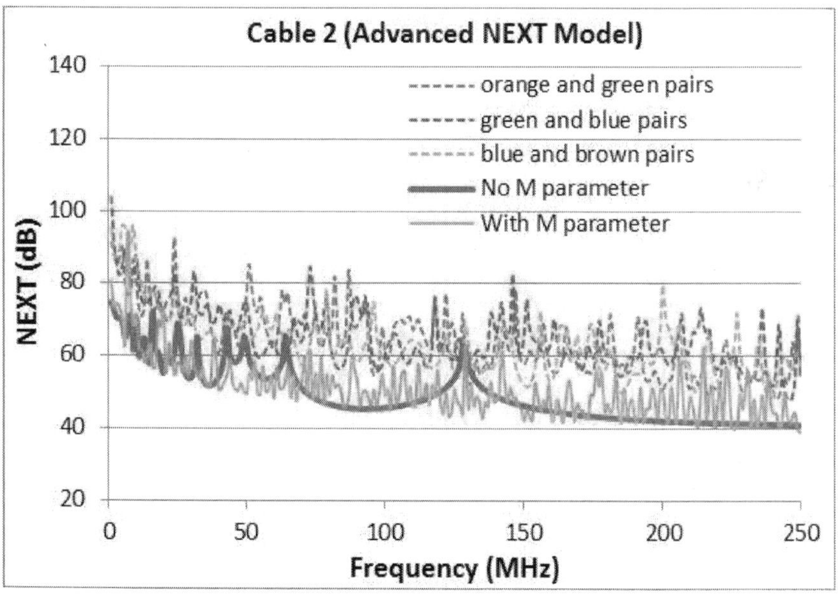

Figure 5.11 Cable 2 NEXT measurements comparison with simulation of the advanced model using the average crosstalk constant with M and no M parameters

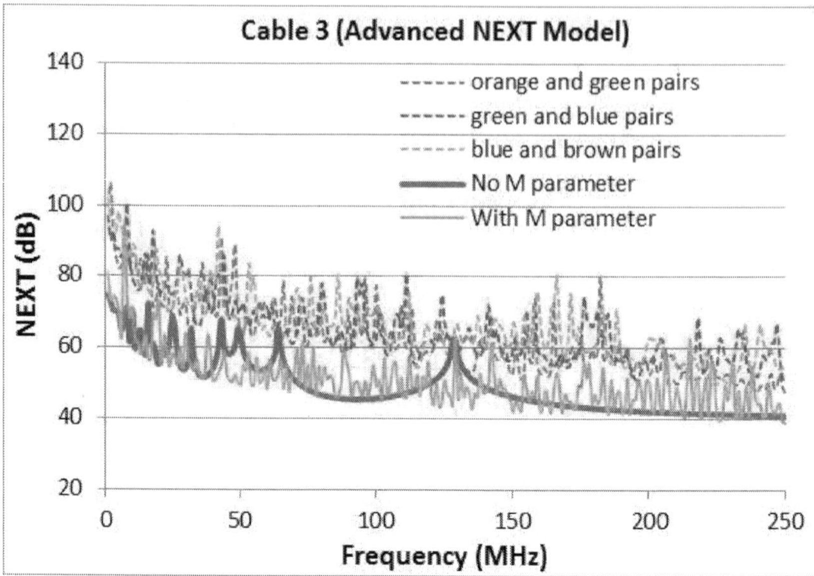

Figure 5.12 Cable 3 NEXT measurements comparison with simulation of the advanced model using the average crosstalk constant with M and no M parameters

The plots of the simulation of the advanced NEXT model in equation (75) with the proposed crosstalk (K_{NEXT}) parameters and M obtained from the average of the curve fitting process in equations (89) and (90) are shown in Figures 5.13 to 5.16 for cables 1 to 3. The plots in Figures 5.13 to 5.16 shows that the use of the proposed crosstalk parameters makes the simulation of the advanced NEXT model to imitate the pattern of the NEXT measurements which is a great improvement from the plots in Figures 5.7 to 5.9 without their use which gave only a smooth curve across the pairs. It was also observed that any increase or decrease in this proposed crosstalk limit shows an under or over prediction of the NEXT measurements showing that is the most acceptable limit across all the pairs of the cables. The FSV GDM comparison of the use of the average crosstalk constant (all cables) in the simulation of the advanced NEXT model in equation (75) is presented in Table 5.2 for cables 1 to 3. The FSV GDM comparison of the proposed crosstalk constant in the simulation of the advanced NEXT model in equation (75) is shown in Table 5.3 for cables 1 to 3.

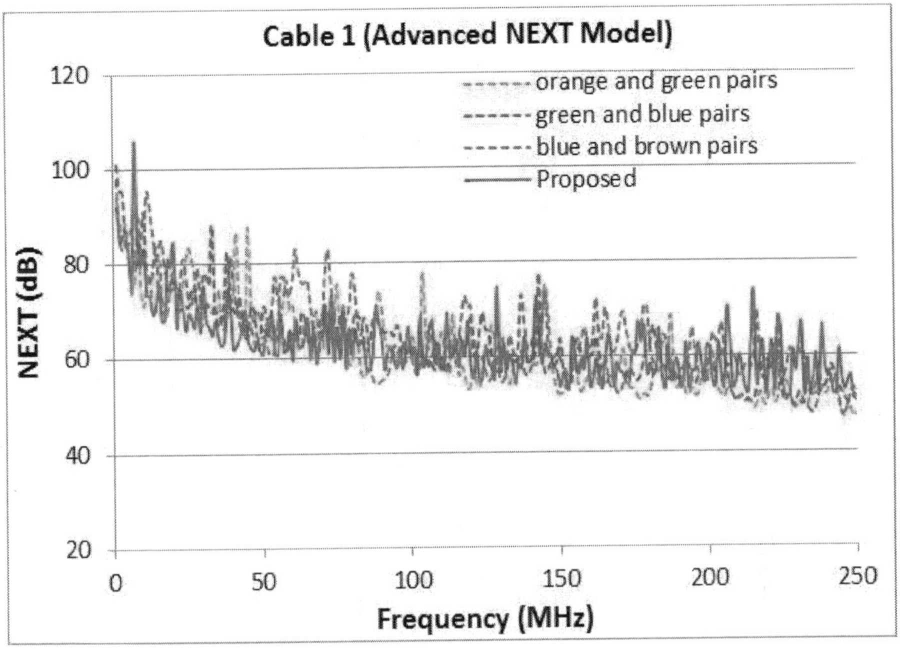

Figure 5.13 Cable 1 NEXT measurements comparison with simulation of the advanced NEXT model using the proposed crosstalk constants

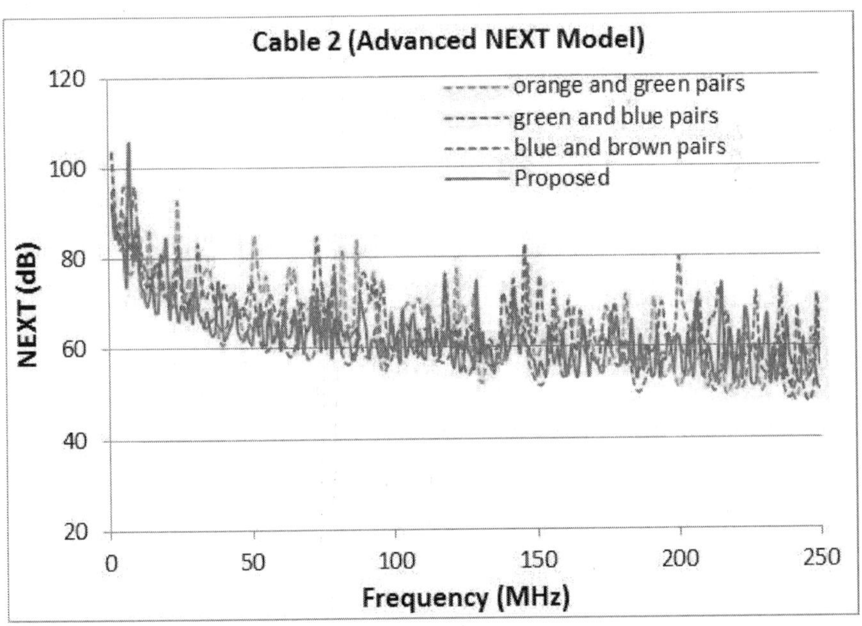

Figure 5.14 Cable 2 NEXT measurements comparison with simulation of the advanced NEXT model using the proposed crosstalk constants

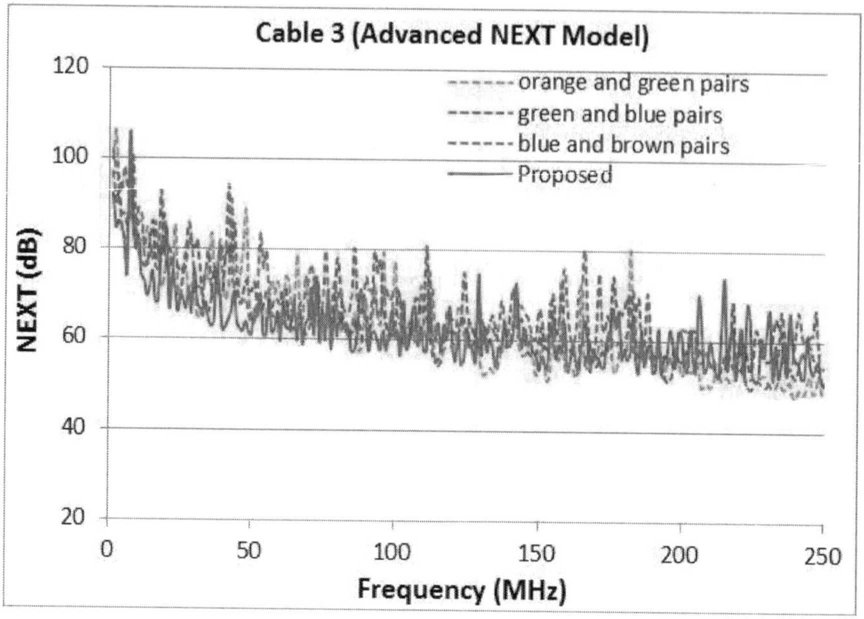

Figure 5.15 Cable 3 NEXT measurements comparison with simulation of the advanced NEXT model using the proposed crosstalk constants

Table 5.2 FSV GDM comparisons of the use of the average K_{NEXT} (all cables) in the simulation of the advanced NEXT model

AVERAGE K_{NEXT} (ALL CABLES)	orange and green pairs	green and blue pairs	blue and brown pairs
CABLE 1	0.8638	0.8353	0.8851
CABLE 2	0.9147	0.9370	0.9136
CABLE 3	0.8158	0.8928	0.8850

Table 5.3 FSV GDM comparisons of the use of the proposed K_{NEXT} in the simulation of the advanced NEXT model

Proposed K_{NEXT}	orange and green pairs	green and blue pairs	blue and brown pairs
CABLE 1	0.5892	0.4865	0.5974
CABLE 2	0.5724	0.5887	0.5312
CABLE 3	0.5749	0.5258	0.5698

The FSV GDM results in Table 5.2 indicates that the use of the average K_{NEXT} (all cables) in the simulation of the advanced NEXT model shows a poor similarity between the NEXT predictions and measurements of all the pair combinations. On the other hand, the use of the improved crosstalk constant results in Table 5.3 shows a fair similarity between predictions and measurements of all the cables pair combinations, which is expected as the crosstalk constant has been found to be unique in each pair combinations [21], [22].

In summary, the single proposed K_{NEXT} for all the cables was obtained from the average of the curve fitting of each pair combination measurements. It was observed during the evaluation that any increase or decrease in this proposed crosstalk constant shows a bad prediction of the NEXT measurements. This shows that is the acceptable crosstalk limit for prediction across all pair combinations of the three cables from different manufacturers. The single proposed crosstalk constant is an improvement on the use of the average crosstalk constant in spite of the diversity in the NEXT measurements of each pair combination. The crosstalk evaluation and prediction method presented can thus be useful to cable engineers investigating crosstalk in data communication systems.

Chapter 6

REVERSE ENGINEERING OF ETHERNET CABLES MEASUREMENT

This chapter presents the method of reverse engineering Ethernet cables impedance profile from return loss measurements using genetic algorithms as explained in section (2.3). These impedance profiles were obtained across the whole cable length using the S-parameters model and genetic algorithms as discussed in section (2.1.5) and (2.3). Four Category 6 UTP cables from different manufacturers were considered for this research to enhance common optimization parameters for the cables. The results were validated with HDTDR impedance profile measurement from the cable analyzer. This method can thus be applied when there is the need for impedance profiles of cables to evaluate their physical integrity or performance before or after installation and only simple (magnitude) tests in the frequency domain are available and time domain tests are inaccessible.

6.1 Reverse Engineering of Ethernet Cables Impedance Profile from Return Loss Measurement

This section presents a technique that can be used to predict the impedance profile of Ethernet cables from return loss measurement at short length intervals using the GA. Four 30m length of Category 6 UTP cables from different manufacturers was considered for this process. The cables are marked as cable 1, CCA cable 2, cable 3 and cable 4. Cables 1, 3 and 4 are common Category 6 UTP cables, while CCA Cable 2 is a Category 6 copper clad aluminum UTP cable. Return loss measurements from two pairs of each cable were considered for the impedance profile extraction process.

The impedance profile was extracted from the return loss measurement by applying the S-parameters expression in equation (25). The GA optimization process involves providing an objective structure and fitness for the GA to use in extracting the best possible impedance profile. The schematic diagram of the objective function is shown in Figure 6.1.

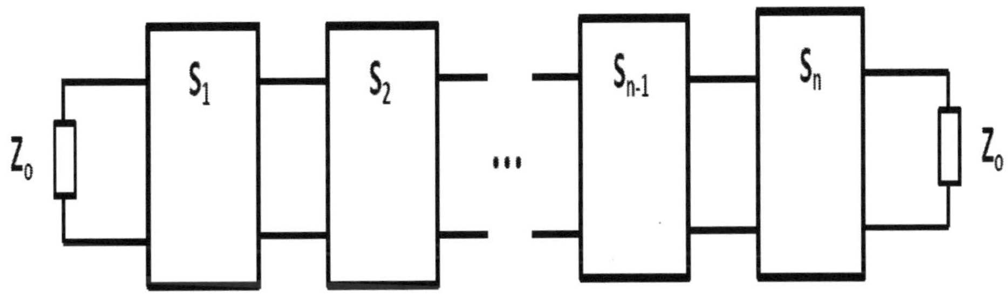

Figure 6.1 Schematic diagram of the cascaded S-parameters for the objective function

The fitness function is designed in such a way as to optimize the new S-parameters model in equation (25) and minimize the differences between predicted and measured return loss data used before extracting the impedance profile. The fitness function is given in equation (91) below and has been used in similar problems in [54], [83].

$$F_{fit} = \frac{1}{n}\sum|S_{11m} - S_{11n}|^2 \quad (91)$$

where, n is number of measured return loss points, S_{11m} is the measured return loss, S_{11n} is the predicted return loss from the new S-parameters model. The flowchart of the GA parameter extraction process is shown in Figure 6.2.

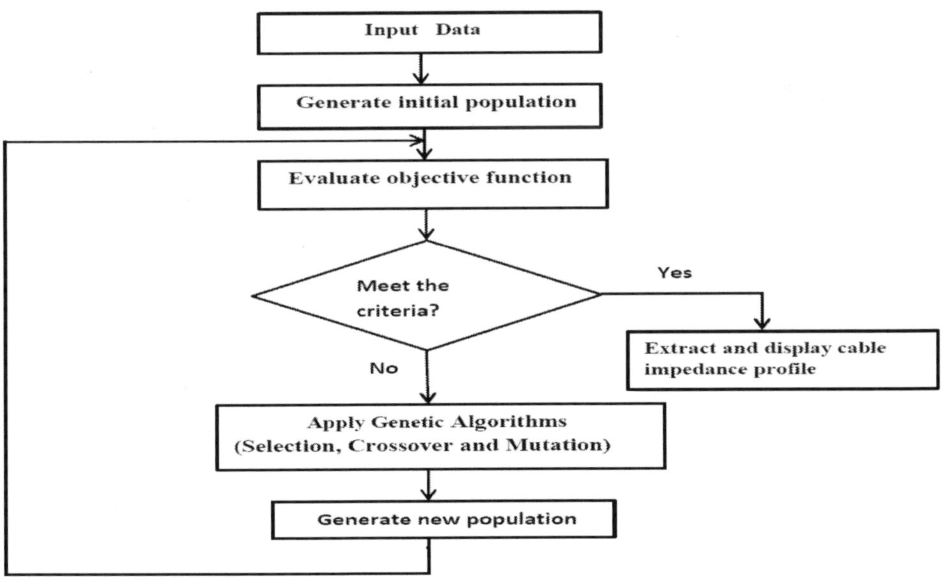

Figure 6.2 The flowchart of the GA impedance profile extraction process

The GA process, its basic implementation and why it was selected in spite of the availability of other optimization algorithms like PSO and SA has been discussed in section (2.3.1.1). The GA optimization process was implemented in MATLAB (Matrix Laboratory) with the starting parameters based on the type of problem and the parameters often used in literature to solve such problems. The starting parameters are: the population size of 233 due to the number of impedances to be extracted, selection: roulette due to explanations [61], [62], crossover: one point, mutation: uniform, mutation rate: 0.001 for a population size of more than 100 as suggested in [61]. For the initial population of the GA, a starting impedance range of 95Ω to 105Ω usually expected from normal cables was selected. The results of the aforementioned parameters gave a maximum MAPE of 1.2167%. However, when the GA operators were changed to mutation: adaptive feasible, crossover: two point and crossover rate: 0.6, the maximum MAPE reduced to 0.9342% showing an improvement using the same impedance range of 95Ω to 105Ω. When the impedance range was reduced to between 97Ω and 103Ω, a further improvement of a maximum MAPE of 0.8486% was obtained. A further reduction of the impedance range from 98Ω to 102Ω further reduced the maximum MAPE to 0.5955%. After the 98Ω to 102Ω impedance range was used, a further reduction in the impedance range did not produce any meaningful results. The graphs of the comparison between measured and extracted impedance profile using the cascaded S-parameters and the GA are shown in Figures 6.3 to 6.6 for cables 1 to 4 using the orange pair. The graphs in Figure 6.3 for cable 1 showed slight differences at some points of no more than 0.5Ω between measured and extracted impedance profile. The graph in Figure 6.4 for cable 2 shows some slight differences at some points with a maximum discrepancy between measured and extracted impedance of about 1.1Ω. The graph in Figure 6.5 for cable 3 also shows slight differences at few points with a maximum discrepancy between measured and extracted impedace profile of about 1.2Ω. The graphs in Figure 6.6 for cable 4 gave a maximum discrepancy of about 0.6Ω. The Mean Absolute Prediction Error (MAPE) of the measured and extracted impedance profile comparison used for the four cables is shown in Table 6.1. The MAPE in Table 6.1 shows that a reasonable and sufficient agreement has been achieved for all the cables considered with a maximum MAPE of 0.5955%.

The graphs of return loss measurement with the S-parameters model simulation using the GA extracted impedance profile are shown in Figures 6.7 to 6.10 for cables 1 to 4. The plot in Figures 6.7 to 6.10 shows a good agreement between measured return loss and S-parameters model simulation using the extracted impedance profile by the GA for all the cables.

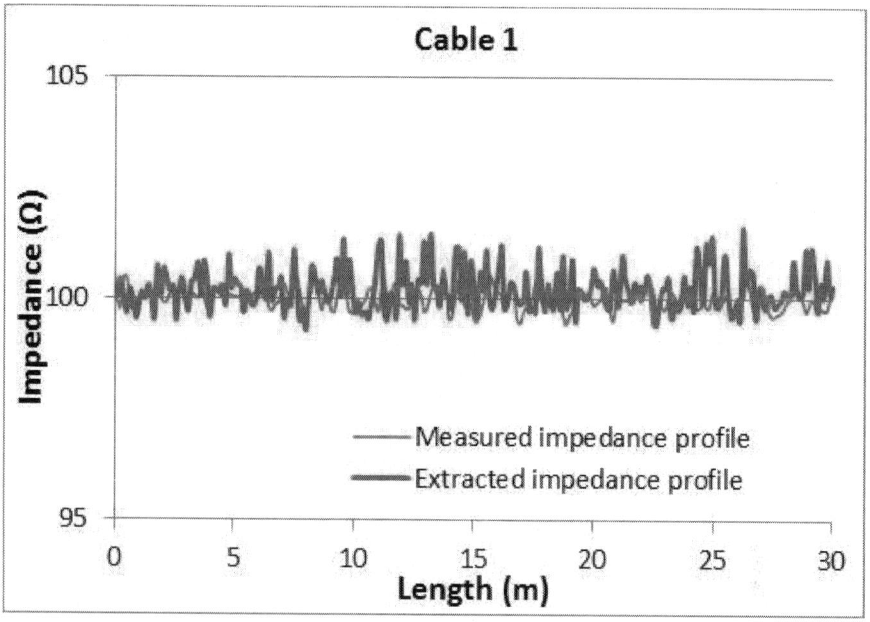

Figure 6.3 Comparison of measured and extracted impedance profile for cable 1

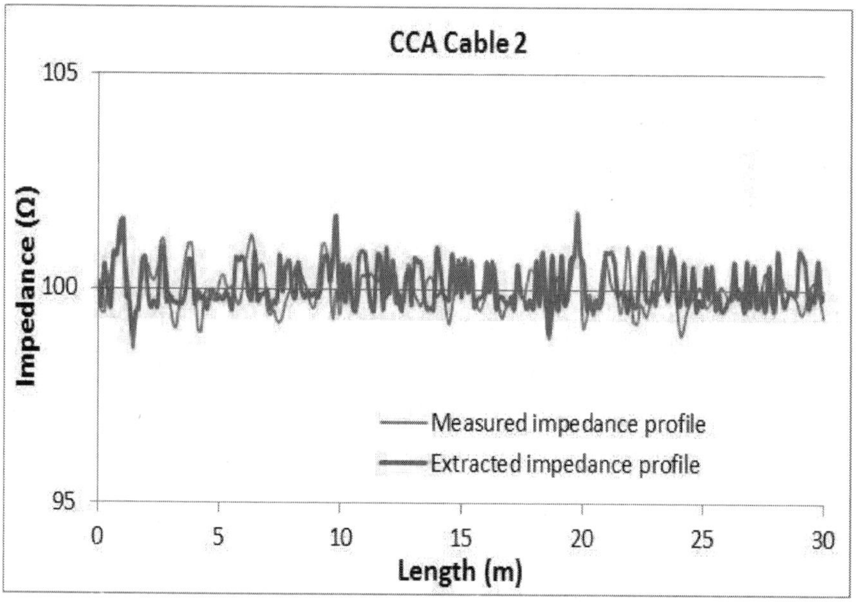

Figure 6.4 Comparison of measured and extracted impedance profile for CCA cable 2

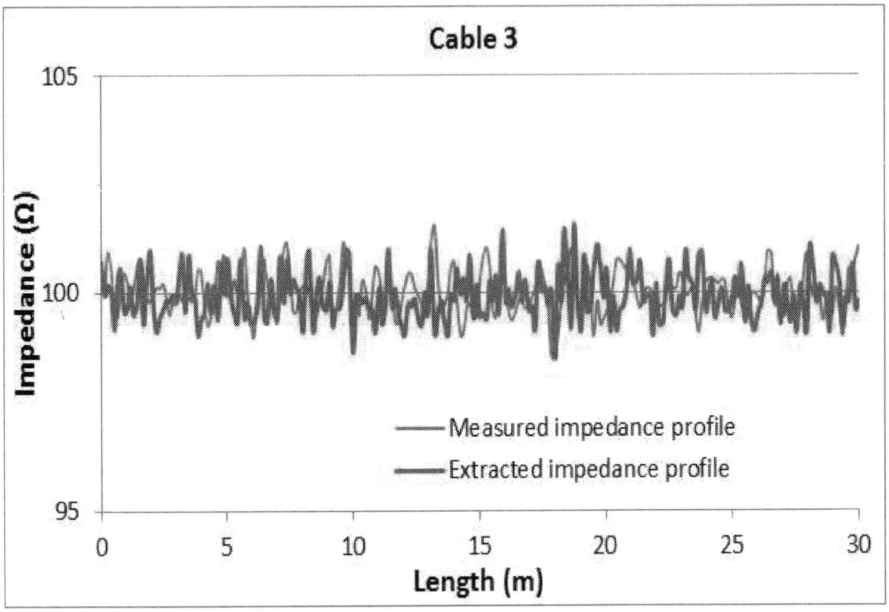

Figure 6.5 Comparison of measured and extracted impedance profile for cable 3

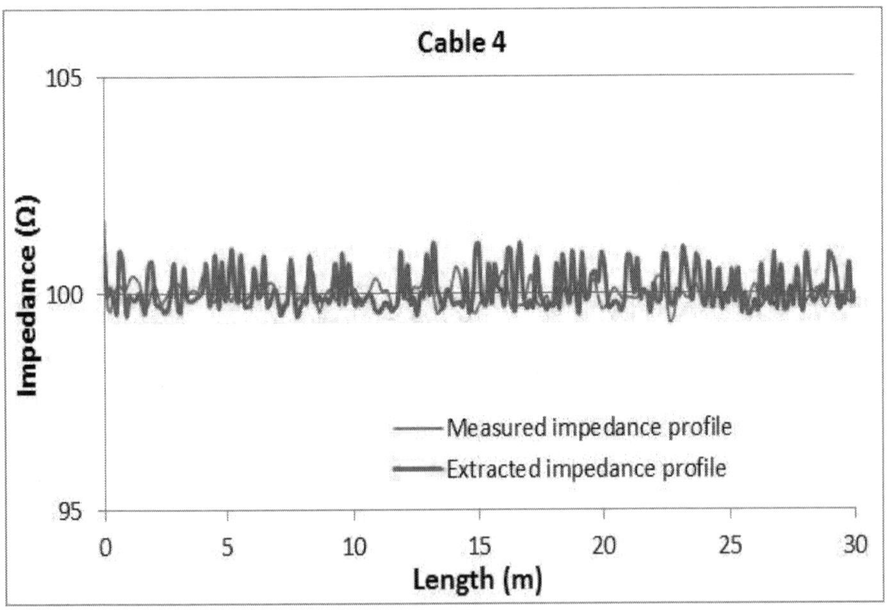

Figure 6.6 Comparison of measured and extracted impedance profile for cable 4

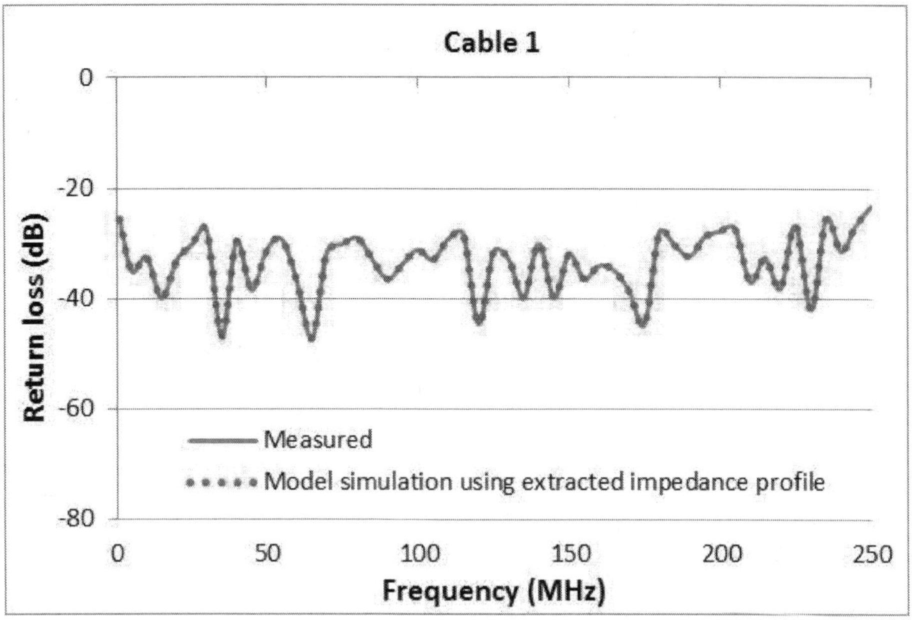

Figure 6.7 Measured and model simulation using the extracted impedance profile for cable 1

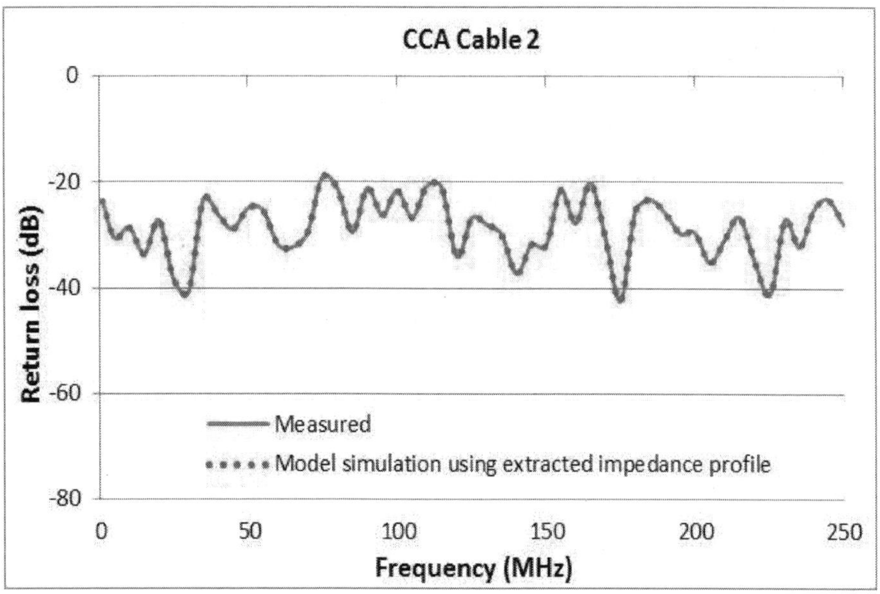

Figure 6.8 Measured and model simulation using the extracted impedance profile for CCA cable 2

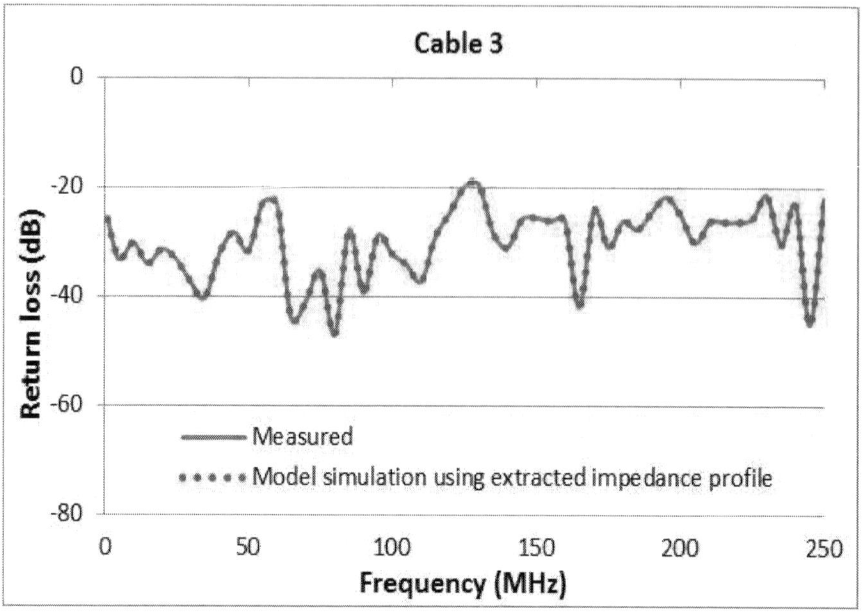

Figure 6.9 Measured and model simulation using the extracted impedance profile for cable 3

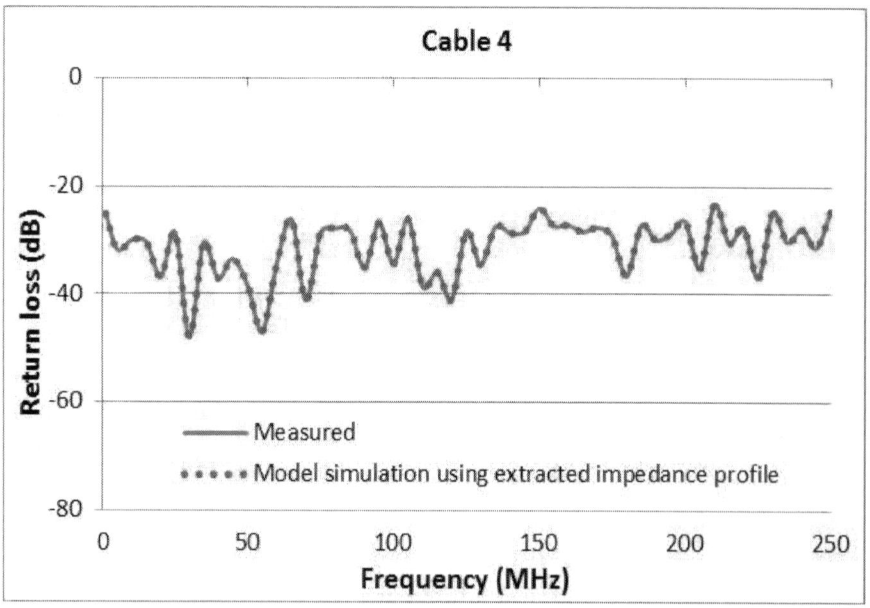

Figure 6.10 Measured and model simulation using the extracted impedance profile for cable 4

Table 6.1 MAPE comparison between measured and extracted impedance profile

CABLES	MAPE
CABLE 1	0.4221%
CCA CABLE 2	0.5799%
CABLE 3	0.5955%
CABLE 4	0.4006%

This section has provided a technique that can be used to reverse engineer Ethernet cables impedance profiles from return loss measurement. The impedance profiles were extracted using cascaded S-parameters and GA. The method can thus be applied in situations where time domain tests are inaccessible or only simple (magnitude) tests in the frequency domain are available and there is the need for impedance profiles of cables to evaluate their performance or physical integrity before or after installation. It is also useful where 'legacy' frequency domain data is the only thing that is available for reference results and there is no upgraded equipment to enable the evaluation of the impedance profiles of the cables.

Chapter 7

PERFORMANCE PARAMETERS PREDICTION IN ETHERNET CHANNELS UNDER STANDARDIZATION

This chapter provides analytical techniques that can be used to predict performance parameters in Ethernet channels under standardization using the 40GBASE-T specifications presented in section (2.2). The analytical method presented is straight forward and is equally applicable to ongoing and future high data rate Ethernet cabling standardization such as the 2.5/5GBASE-T and 50/100GBASE-T. The method can thus be used by cable designers in virtual tests during the standardization process and prototype design.

7.1 40GBASE-T Channel Insertion Loss Modeling

The 40GBASE-T channel insertion loss was modeled using the S-parameters method presented in section (2.1.4) and the insertion loss specifications in section (2.2.1). The two channels selected for modeling are the 3m-24m-3m and 1m-10m-1m as specified by the 40GBASE-T Task Force in [35]. The 3m and the 1m are the patch cord lengths, while 24m and 10m are the cable lengths. Cable and patch cord asymptotic impedance of 104.5Ω and 95.5Ω were considered as given in [35] for the 40GBASE-T channel modeling. The insertion loss modeling results was validated with the 40GBASE-T channel predictions. The S-parameters expression used to determine the insertion loss and return loss has been explained in section (2.1.4) and is given as:

$$S = \frac{1}{K_s}\begin{bmatrix}(Z_{ck}^2 - Z_r^2)\sinh(\gamma_k l) & 2Z_{ck}Z_r \\ 2Z_{ck}Z_r & (Z_{ck}^2 - Z_r^2)\sinh(\gamma_k l)\end{bmatrix} \tag{92}$$

where, Z_{ck} and Z_r are the impedance of the cable or patch cord and reference impedance of the measurement equipment respectively in ohms, l is the length of the cable in meters and γ_k is the propagation constant.

$$K_s = 2Z_{ck}Z_r\cosh(\gamma_k l) + (Z_{ck}^2 + Z_r^2)\sinh(\gamma_k l) \tag{93}$$

$$Z_{ck} = Z_a \left[1 + 0.055 \frac{(1-j)}{\sqrt{f_k}}\right] \quad \text{(ohms)} \tag{94}$$

Z_a is the asymptotic impedance of the cable.

The comparison of the insertion loss using the S-parameters and the 40GBASE-T predictions are presented in Figure 7.1 and Figure 7.2. The graphs in Figure 7.1 and Figure 7.2 show good agreement between the S-parameters method and the 40GBASE-T. The modeling results obtained indicates that further study can be conducted with this method.

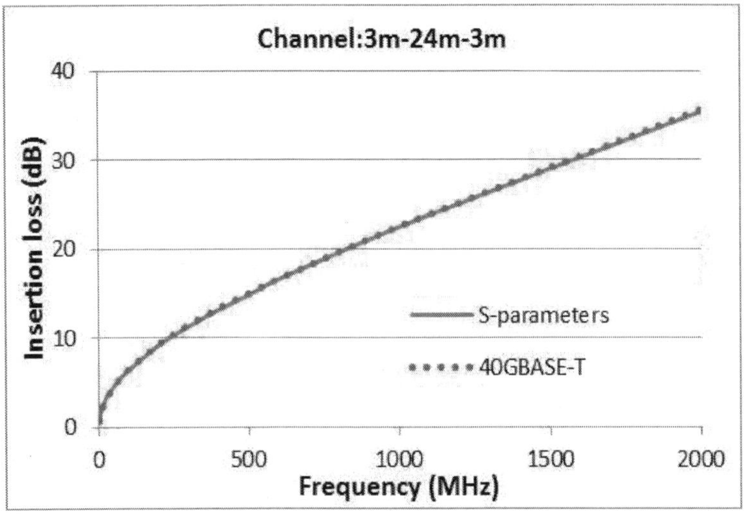

Figure 7.1 3m-24m-3m channel insertion loss comparison

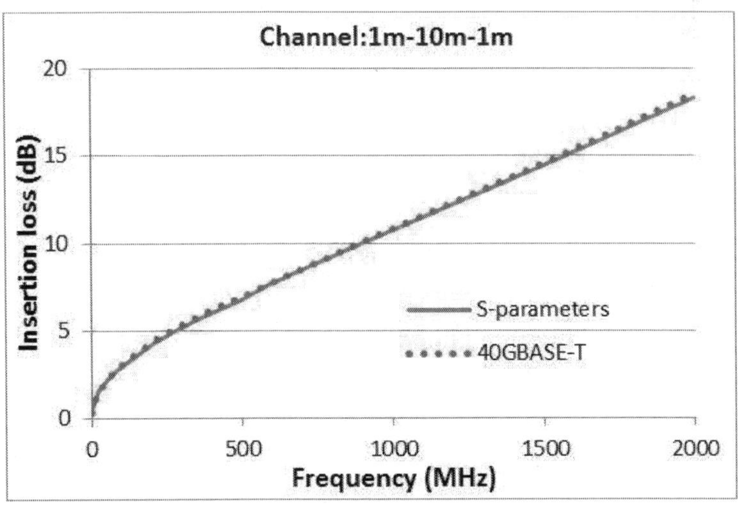

Figure 7.2 1m-10m-1m channel insertion loss comparison

7.2 40GBASE-T Channel Return Loss Modeling

The 40GBASE-T channel return loss was also modeled using the S-parameters method in section (2.1.4) and the 40GBASE-T in section (2.2.1) following the process as explained in Section (7.1) for the 40GABSE-T insertion loss modeling. Cable and patch cord asymptotic impedance of 104.5 ohms and 95.5 ohms were considered as given in [35] for the 40GBASE-T channel modeling. The plots of the comparison between the S-parameters method and the 40GBASE-T are shown in Figure 7.3 and Figure 7.4 for 3m-24m-3m and 1m-10m-1m channels respectively. The graphs in Figure 7.3 and 7.4 show good agreements with the S-parameters method.

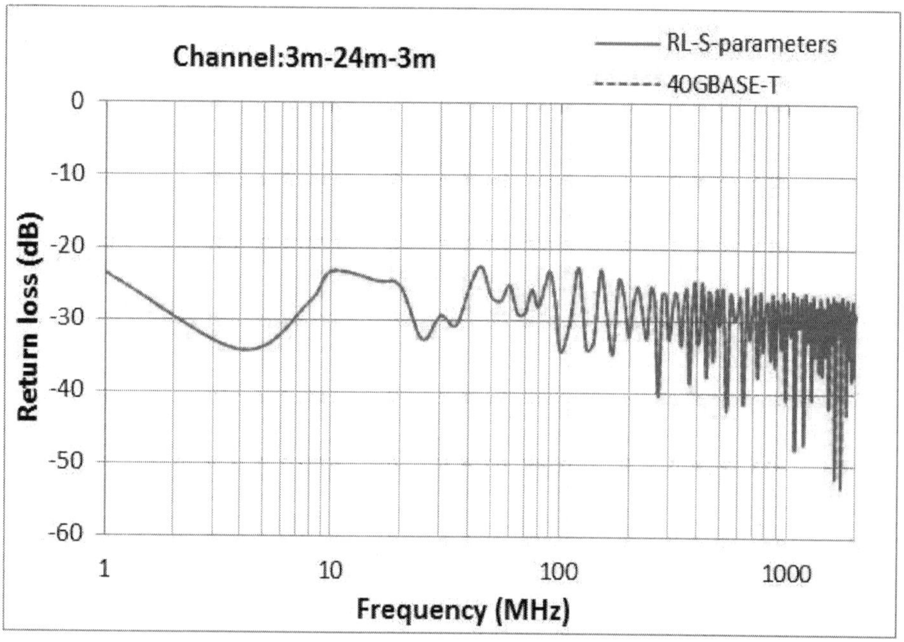

Figure 7.3 3m-24m-3m channel return loss comparison

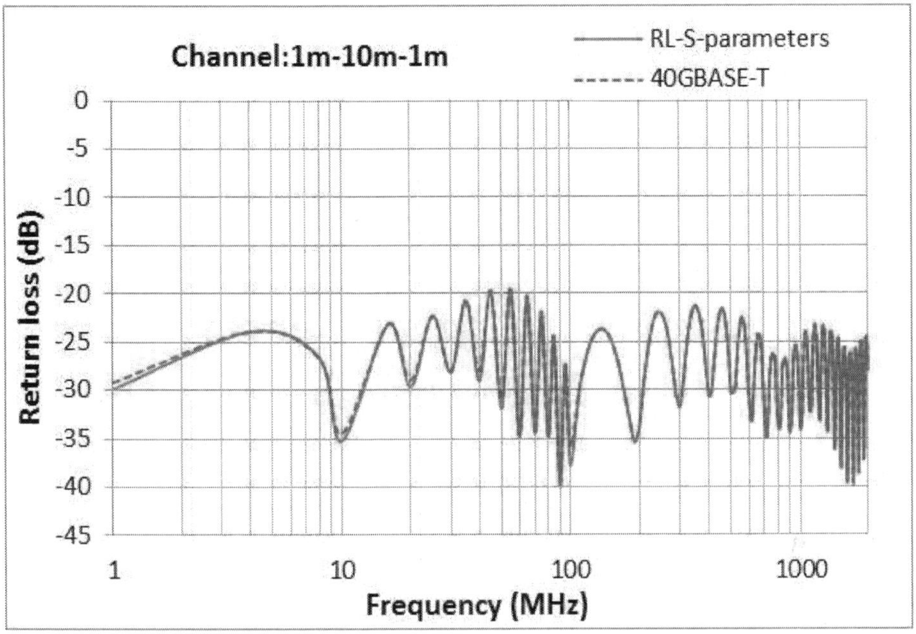

Figure 7.4 1m-10m-1m channel return loss comparison

The significance of the above research is that it provides a method that can be used to predict the insertion loss and return loss of different channel configurations allowed by the 40GBASE-T. The research provides engineers with tools to investigate channel behavior under different configurations which could be extended to other high data rate cabling and standardization.

7.3 Effects of Defects on the Return Loss of 40GBASE-T Channel Configurations

This section provides a guided simulation method that can be used to predict the effects of defects on two different 40GBASE-T channel configurations using periodic impedance changes. Cable and patch cords asymptotic impedance of 104.5 ohms and 95.5 ohms was considered for use as explained in section (2.1.4) and (2.2.1). The channel configurations considered for this research are: 3m-24m-3m and 1m-10m-1m channels as given in [35]. The 3m-24m-3m channel has two patch cords of length 3m each, with a backbone cable of 24m, while the 1m-10m-1m channel has two patch cords of length 1m each, with a backbone cable of 10m. The channels were simulated using periodic impedance changes of +5%/-5%, +10%/-10% and +15%/-15% over 100 cascaded S-parameters or segments of the backbone cable. This means that the periodic impedance changes will be at 0.24m per segment for the 3m-24m-3m channel, while the periodic impedance changes will be at 0.1m per segment for the 1m-10m-1m channel. The method was used to predict the effects of the maximum and minimum impedance tolerance levels often specified for Ethernet cables [96] as +15%/-15% of 100Ω on return loss. However, the example used in this research for illustration is an extreme case test of 100 periodic impedance changes along the whole length to be conducted in order to mimic the highest degree of impedance variations within the tolerance levels that could occur after installation.

The plots of the return loss comparison for the 3m-24m-3m and 1m-10m-1m channels using impedance variations with the 40GBASE-T limits [96] are shown in Figures 7.5 and 7.6 respectively. An analysis of the plot in Figure 7.5 for 3m-24m-3m channel shows that the return loss from the +15%/-15% crossed the 40GBASE-T limit at about 170MHz, followed by the +10%/-10% and the +5%/-5% impedance changes at about 180MHz and 200MHz respectively. Similarly, the graph in Figure 7.6 for the 1m-10m-1m channel shows that the return loss from the +15%/-15% was the first to cross the limit at about 370MHz, followed by the +10%/-10% impedance change at about 440MHz. The return loss from the +5%/-5% impedance change shows that it crosses the limit at about 480MHz.

The summary of the result is that for both the 3m-24m-3m and 1m-10m-1m channels, the 15%/-15% percentage impedance change was the first to cross the limit followed by the other lower percentage impedance changes. The results of the research also showed that the return loss from the longer channel (3m-24m-3m) first crossed the limit at about 170MHz,while the shorter channel (1m-10-1m) first crossed the limit at about 370MHz. However, this is an extreme case test (100 periodic impedance changes) of 5% to 15% along the whole cable length which is not expected in most normal network situations. It was conducted to mimic the highest degree of impedance variations within the tolerance levels that could occur after installation. The method can be used by cable designers in virtual stress tests during the cable prototype design before their manufacture.

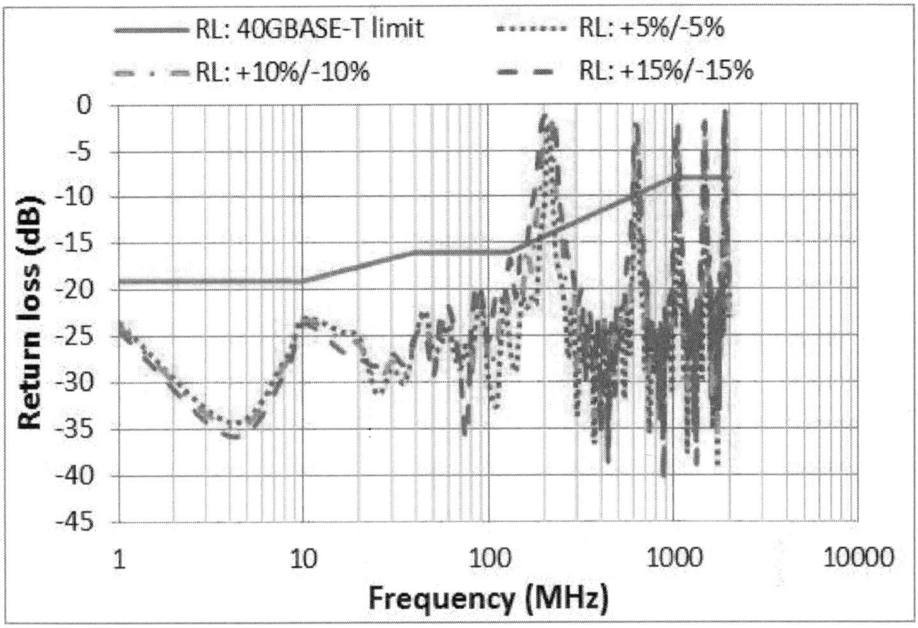

Figure 7.5 3m-24m-3m channel return loss prediction using periodic impedance variations

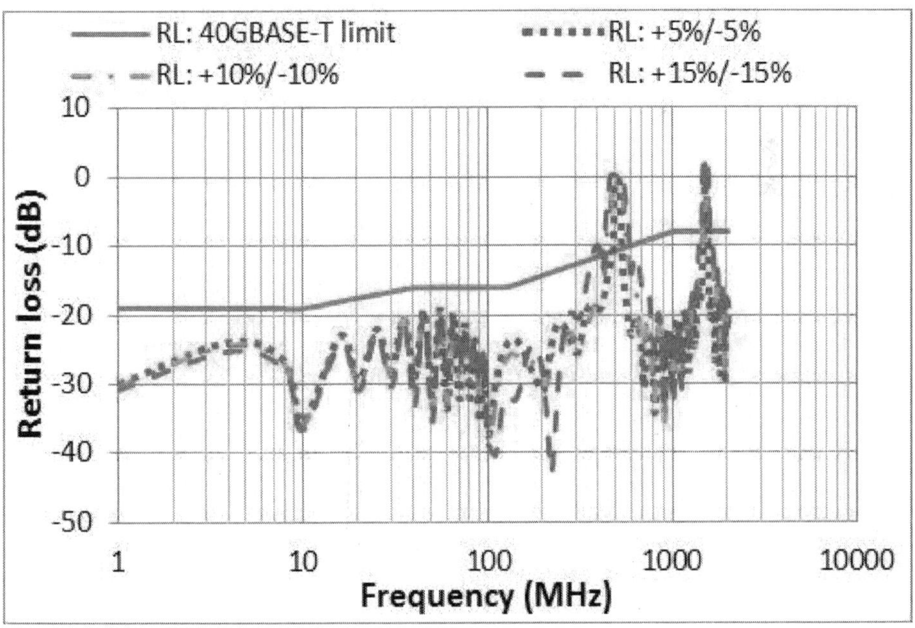

Figure 7.6 1m-10m-1m channel return loss prediction using periodic impedance variations

This section has provided an analytical technique that can be used to predict performance parameters such as return loss using periodic impedance variations that mimic the effects of defects on Ethernet channels using the maximum and minimum impedance tolerance levels often specified for them. The analytical method can thus be used by cable designers in virtual tests during the standardization process and prototype design. This technique will be also useful to cable engineers investigating performance parameters in ongoing and future Ethernet channels standardization research.

Chapter 8

CONCLUSION AND FUTURE WORK

This chapter provides the conclusion to the research in this thesis and suggestions for future work.

8.1 Research Summary and Conclusion

The thesis provided a technique that can be used by cable professionals, installers and engineers to assess cable-key performance measurements before deployment that does not rely heavily on human subjective judgement and enables objective decisions in the choice of cables. The assessment method included subjecting four Category 6 UTP cables from different manufacturers to three rounds of coiling and stretching out tests to mimic the highest degree of handling effects situations that could occur during or after installation. This method can help engineers/installers determine which of the cables has the highest resilience or otherwise to handling stress. The method can also help engineers monitor variations in key performance parameters which could not have been easy to do with the human eye.

The research provided a technique that can be used to evaluate and model NEXT in UTP cables using the standard simple and advanced models. The research evaluated the ANSI and average crosstalk constants of three UTP cables from different manufacturers. The result of this analysis was used to provide improved crosstalk constants for fast NEXT predictions in Category 6 UTP cables. The NEXT evaluation and modeling method presented in this thesis can be used by cable engineers investigating NEXT in the design of data communication systems.

A technique that can be used to reverse engineer Ethernet cables impedance profiles from return loss measurement was also provided in this thesis report. The impedance profiles were extracted using cascaded S-parameters and genetic algorithms. This method is applicable in situations where time domain tests are inaccessible or only simple (magnitude) tests in the frequency domain are available and there is the need for impedance profiles of cables to evaluate their performance or physical integrity before or after installation. It is also useful where 'legacy' frequency domain data is the only thing that is available for reference results and there is no upgraded equipment to enable the evaluation of the impedance profiles of the cables.

Finally, guided simulation technique that can be used to predict key performance parameters in Ethernet channels under standardization was provided using cascaded S-parameters model. The 40GBASE-T was used as a sample to predict the return loss using specified impedance variations. The method can be used by cable designers in virtual tests during the cable prototype design before their manufacture. The guided simulation method is equally applicable to ongoing and future high data cabling and standardization such as 2.5/5 GBASE-T and 50/100GBASE-T.

In conclusion, the research has meet all its aims and objectives by providing a set of analytical and forensic tools for Ethernet cabling to assist designers and those involved in deployment in analyzing cable performance and the reasons behind the actual performance obtained.

8.2 Future Work

Future work can be on investigating other methods of assessing cables key performance measurements in order to meet the high quality network demands of the current IOT infrastructures in the mist of counterfeit and nonstandard compliant cables in the market. Extending the crosstalk evaluation and modeling method presented in this thesis report to higher categories of cables such as Category 7 and even Category 8 cables depending on the availability of test equipment will also be of value to engineers investigating crosstalk in data communication systems.

Future work in extending the cascaded S-parameters method applied in the 40GBASE-T performance parameters prediction to ongoing and future high data cabling and standardization such as 2.5/5 GBASE-T and 50/100GBASE-T will also be of value to the body of knowledge.

References

[1] D.Valencic, V.Lebinac and A.Skendzic, "Developments and Current Trends in Ethernet Technology," in *36th International Convention on Information and Communication Technology,* Opatija, 2013.

[2] K. Azadet, "Gigabit Ethernet over Unshielded Twisted Pair Cables," in *International Symposium on VLSI Technology,Systems and Applications,* Taipei, 1999.

[3] C.Spurgeon and J.Zimmerman, Ethernet: The Definitive Guide, Second Edition, O'Reilly Media Inc., 2014.

[4] H. Lajmi, M. Alimi and S.Ajili, "Using Ethernet Technology for In-Vehicle's Network Analysis," in *Fifth International Conference on Computational Intelligence, Communication Systems and Networks,* Madrid, 2013.

[5] D.Law, D.Dove, J.D'Ambrosia, M.Hajduczenia, M.Laubach and S.Carlson, "Evolution of Ethernet Standards in the IEEE 802.3 Working group," *IEEE Communications Magazine,* vol. 51, no. 8, pp. 88-96, 2013.

[6] F. Straka, "40 Gbits/s over twisted-pair copper cable is on the way," *Electronics Products Magazine,* vol. 56, no. 4, pp. 20-22, 2013.

[7] P. McLaughlin, "Category 8 questions and answers," *Cabling Installation and Maintenance,* vol. 23, no. 10, pp. 23-25, 2015.

[8] O. Savi, "10G Ethernet Over Structured Copper Cabling," Siemen, [Online]. Available: https://www.siemon.com/us/white_papers/04-12-22_10G_Ethernet_Over_Structured_Copper_Cabling.asp. [Accessed 30 August 2016].

[9] S.Lampen, M. Burgt and C.Dole, "High-Definition Cabling and Return Loss," *SMPTE,* vol. 110, no. 1, pp. 34-38, 2001.

[10] J.Jin, J.Gubbi, S.Marusic and M.Palaniswami, "An Information Framework for Creating Smart City through the Internet of Things," *IEEE Internet of Things,* vol. 1, no. 2, pp. 112-121, 2014.

[11] R.Harmon, C.Leon and S.Bhide, "Smart Cities and the Internet of Things," in *International Conference on Management of Engineering and Technology*, Portland, 2015.

[12] Fluke Networks, "Application Note: Copper Clad Aluminum (CCA) Cables," Fluke Networks, [Online]. Available: http://www.flukenetworks.com/doc_links_pdf/en/content/application-note-copper-clad-aluminum-cables. [Accessed 30 August 2016].

[13] D. Kiddoo, "Update: Counterfeit and Non-Compliant Communications Cable," in *63rd International Wire and Cable Symposium*, Rhode Island, 2014.

[14] P.McLaughlin, "Counterfeit cable is getting ugly," *Cabling Installation and Maintenance*, vol. 19, no. 8, 2011.

[15] P. McLaughlin, "Copper-clad aluminium conductors the latest counterfeiting ploy," *Cabling Installation and Maintenance*, vol. 19, no. 4, 2011.

[16] F.Peri, "Non-Compliant and Counterfeit Cable:A Risk Too Real to Ignore," [Online]. Available: http://cccassoc.org/files/6314/0050/5728/CCCA-Article-on-Non-Compliant-Cable-in-ICT-Today-May-2014.pdf. [Accessed 30 August 2016].

[17] Standard IEEE, *IEEE 1597.1 Standard for Validation of Computational Electromagnetics,Computer Modeling and Simulations*, Standard IEEE, 2009.

[18] F. Massey, "The Kolmogorov-Smirnov Test for Goodness of Fit," *Journal of the American Statistical Association*, vol. 46, no. 253, pp. 68-78, 1951.

[19] S.Sivanandam and S.Deepa, Introduction to Genetic Algorithms, Springer, 2008.

[20] S. Naimi, B. Hajji, Y. Habbani, I. Humenyuk, J. Launay and P. Boyer, "Modeling of the pH-ChemFET response and using Genetic Algorithms as extraction parameters method," in *Faible Tension Faible Consommation (FTFC)*, Marrakech, 2011.

[21] P.Lafata and M.Pravda, "Analyzing and Modeling of Far-End Crosstalk in Twisted Multi-Pair Metallic Cables," in *International Conference on Applied Electronics*, Pilsen, 2011.

[22] P.Latafa, "Realistic Modeling of Far-End Crosstalk in Metallic Cables," *IEEE Communication Letters*, vol. 17, no. 3, pp. 435-438, 2013.

[23] ANSI Standard, *A new analytical method for NEXT and FEXT noise calculation,* Huntsville: ANSI Standard T1E1.4, 1998.

[24] A.Zadehgol, "An efficient approximation for arbitrary port suppression of multiport scattering parameters," *International Journal of Numerical Modeling: Electronic Networks, Devices and Fields,* vol. 27, no. 1, pp. 164-172, 2014.

[25] M.Hua, Y. Liang, T.Anju and W. Zhaojin, "The Design and Simulation of Signal Integrity for High-Speed Backplane based on VPX," in *10th IEEE Conference on Industrial Electronics and Applications (ICIEA),* Auckland, 2015.

[26] S.Muller, T. Reuschel, R.Donadio, Y. Kwark, H. Bruns and C. Schuster, "Energy-Aware Signal Integrity Analysis for High-Speed PCB Links," *IEEE Transactions on Electromagnetic Compatibility,* vol. 57, no. 5, pp. 1226-1234, 2015.

[27] J. Choma and W.Chen, Feedback Networks: Theory and Circuit Applications, World Scientific Publishing Company, 2007.

[28] R.Mavaddat, Network Scattering Parameters, World Scientific Company, 1996.

[29] R. Papazyan, P. Pettersson, H. Edin, R. Eriksson and U. Gafvert, "Extraction of High Frequency Power Cable Characteristics from S-parameter Measurements," *IEEE Transactions on Dielectrics and Electrical Insulation,* vol. 2, no. 3, pp. 461-470, 2004.

[30] E.Silva, High Frequency and Microwave Engineering, Butterworth Heinemann, 2001.

[31] J. Dobrowolski, Microwave Network Design Using the Scattering Matrix, Artech House, 1981.

[32] C.Poole and I.Darwazeh, Microwave Active Circuit Analysis and Design, Elsevier, 2015.

[33] A.Khan, Microwave Engineering: Concepts and Fundamentals, Taylor & Francis Group, 2014.

[34] K. Gupta, R. Garg and R.Chadha, Computer Aided Design of Microwave Circuits, Artech House, 1981.

[35] P.Kish, *Channel Return loss Modeling Results,* IEEE 802.3bq adhoc Standards Contributions, 2013.

[36] V.Zdorov, K. Doshi and P.Pupalaikis, "Computation of Time Domain Impedance Profile S-Parameters: Challenges and Methods," in *Design Conference (Design Con 2014)*, Santa Clara, 2014.

[37] D.DeGroot, K.Doshi, D.Dunham and P.Pupalaikis, "De-embedding in High Speed Design," in *Design Conference (Design Con 2012)*, Santa Clara, 2012.

[38] D. Pozar, Microwave Engineering,Fourth Edition, John Wiley and Sons Inc., 2012.

[39] F.Ulaby, E.Michielssen and U.Ravaioli, Fundamentals of Applied Electromagnetics, 6th Edition, Prentice Hall, 2010.

[40] M.Degerstrom, B.Gilbert and E. Daniel, "Accurate resistance, inductance, capacitance and conductance (RLCG) from uniform transmission line measurements," in *IEEE-EPEP Electrical Performance of Electronic Packaging Conference*, San Jose, 2008.

[41] M.Sampath, "On Addressing the Practical Issues in the Extraction of RLGC Parameters for Lossy Multiconductor Transmission Lines using S-parameter Models," in *IEEE-EPEP Electrical Performance of Electronic Packaging*, San Jose, 2008.

[42] C.DiMinico, *40GBASE-T PHY-Channel Insertion Loss*, IEEE 802.3bq adhoc 40GBASE-T Channel Modeling Standards Contributions, 2013.

[43] O. Ogundapo, A.Duffy and C.Nche, "Insertion and Return Loss Modeling for 40GBASE-T Channel Configurations," in *30th Annual Review in Applied Computational Electromagnetics*, Jacksonville, 2014.

[44] S.Rao, Engineering Optimization: Theory and Practice, Fourth Edition, John Wiley and Sons, Inc., 2009.

[45] M. Tabares, J. Garzon and C. Dominguez, "Multiobjective Optimization-based Design of Wearable Electrocardiogram Monitoring Systems," in *36th IEEE Annual International Conference on Engineering in Medicine and Biology Society*, Chicago, 2014.

[46] X. Sheng, S. Johnson, J.Michel and L. Kimerling, "Optimization-Based Design of Surface Textures for Thin-Film Si Solar Cells-Are Conventional Lambertian Models Relevant?," in *37th IEEE Conference on Photovoltaic Specialists Conference*, Seattle, 2011.

[47] R.Singh, A. Gole, P. Graham, S. Filizadeh, C.Muller and R.Jayasinghe, "Grid Processing for Optimization Based Design of Power Electronic Equipment using Electromagnetic Transient Simulation," in *25th IEEE Canadian Conference on Electronics and Computer Engineering*, Quebec, 2012.

[48] W.Moutassem and G.Anders, "Configuration Optimization of Underground Cables for Best Ampacity," *IEEEE Transactions on Power Delivery,* vol. 25, no. 4, pp. 2037-2045, 2010.

[49] K. Deb, Optimization for Engineering Design: Algorithms and Examples, Second Edition, Prentice-Hall Ltd, 2012.

[50] E.Eilam, Reversing: Secrets of Reverse Engineering, First Edition, John Wiley and Sons Ltd., 2005.

[51] A. Telea, Reverse Engineering-Recent Advances and Applications, InTech, 2012.

[52] W.Zhang, K.Wang and J.Meng, "Research on Application of Functional FMECA in Reverse Engineering Optimization," in *First International Conference on Reliability Systems Engineering*, China, 2015.

[53] K.Krogmann, M. Kuperberg and R. Reussner, "Using Genetic Search for Reverse Engineering of Parametric Behavior Models for Performance Prediction," *IEEE Transactions on Software Engineering,* vol. 36, no. 6, 2010.

[54] E.Gadjeva, V.Durev and M.Hristov, Analysis, Model Parameter Extraction and Optimization of Planar Inductors using MATLAB ; Matlab-Modelling, Programming and Simulations, Emilson Pereira Leite (Ed.), InTech, 2010.

[55] C.Balubal, A. Bernado, B. Lasheras, R. Uyehara, A.Bandala and E.Dadios, "Cabling and Cost Optimization System for IP Based Networks through Genetic Algorithms," in *IEEE Region 10 Symposium*, Malaysia, 2014.

[56] V. Lute, A.Upadhyay and K.Singh, "Genetic Algorithms-Based Optimization of Stayed Bridges," *Journal of Software Engineering and Applications*, vol. 4, no. 10, pp. 571-578, 2011.

[57] N.Moldovan, R. Picos and G.Moreno, "Parameter Extraction of a Solar Cell Compact Model using Genetic Algorithms," in *Proceedings of the Spanish Conference on Electron Devices*, Santiago de Compostela, 2009.

[58] A. Harrag and S. Messalti, "Extraction of solar cell parameters using genetic algorithm," in *4th International Conference on Electrical Engineering (ICEE)*, Boumerdes, 2015.

[59] J. Watts, C.Bittner, D.Heaberlin and J. Hoffmann, "Extraction of Compact Model Parameters for ULSI MOSFETs Using a Genetic Algorithm," in *International Conference on Modeling and Simulation of Microsystems*, Puerto Rico, 1999.

[60] G.Liu and J.Chen, "The Application of Genetic Algorithm Based Matlab in Function Optimization," in *International Conference on Electrical and Control Engineering*, Yichang, 2011.

[61] K.F.Man, K.S.Tang and S.Kwong, "Genetic Algorithms:Concepts and Applications," *IEEE Transactions on Industrial Electronics*, vol. 43, no. 5, pp. 519-534, 1996.

[62] A.Shukla, H.Pandey and D.Mehrotra, "Comparative Review of Selection Techniques in Genetic Algorithm," in *1st International Conference on Futuristic trend in Computational Analysis and Knowledge Management*, Greater Noida, 2015.

[63] N. Razali and J. Geraghty, "Genetic Algorithm Performance with Different Selection Strategies in Solving TSP," in *Proceedings of the World Congress on Engineering*, London, 2011.

[64] A. Cuellar, R.R.Troncoso, L.M.Velazquez and R.Rios, "PID-Controller Tuning Optimization with Genetic Algorithms in Servo Systems," *International Journal of Advanced Robotics,Intechopen*, vol. 10, no. 2, pp. 1-14, 2013.

[65] F. Alabsi and R. Naoum, "Comparison of Selection Methods and Crossover Operations using Steady State Genetic Based Intrusion Detection System," *Journal of Emerging Trends in Computing and Information Sciences*, vol. 3, no. 7, pp. 1053-1058, 2012.

[66] J. Mendes and A. Almeida, "A Comparative Study of Crossover Operators for Genetic Algorithms to Solve the Job Shop Scheduling Problem," *WSEAS Transactions on Computers*, vol. 12, no. 4, pp. 164-173, 2013.

[67] N. Soni and T. Kumar, "Study of Various Crossover Operators in Genetic Algorithms," *International Journal of Computer Science and Information Technologies*, vol. 5, no. 6, pp. 7235-7238, 2014.

[68] N. Soni and T. Kumar, "Study of Various Mutation Operators in Genetic Algorithms," *International Journal of Computer Science and Information Technologies*, vol. 5, no. 3, pp. 4519-4521, 2014.

[69] T.Laseetha and R.Sukanesh, "Investigation on the Performance of Linear Antenna Array synthesis using Genetic Algorithm," *Journal of Selected Areas in Telecommunications (JSAT),* vol. 1, no. 2, pp. 60-66, 2011.

[70] C.Fyfe, P.Tino, D.Charles, C. Osorio and H.Yin, Intelligent Data Engineering and Automated Learning -IDEAL 2010, Spinger, 2010.

[71] Y. Sanchez, C.Rosi, E.Paez and M.Azpurua, "Feature Selective Validation Applied to the Comparison of Calibration Data," in *International Conference on Precision Electromagnetic Measurements,* Rio de Janeiro, 2014.

[72] A.Duffy, A.Orlandi and H.Sasse, "Offset Difference Measure Enhancement for the Feature Selective Validation Method," *IEEE Transactions on Electromagnetic Compatibility,* vol. 50, no. 2, pp. 413-415, 2008.

[73] V. Rajamani, C.Bunting, A.Orlandi and A.Duffy, "Introduction to Feature Selective Validation (FSV) Method," in *International Symposium on Antennas and Propagation,* Albuquerque, 2006.

[74] O.Ventosa, M.Pous, F.Silva and R.Jauregui, "Application of Feature Selective Validation Method to Pattern Recognition," *IEEE Transactions on Electromagnetic Compatibility,* vol. 56, no. 4, 2014.

[75] A.Duffy, H.Sasse, B. Archambeault and A.Drozd, "Overview and Update on IEEE Std.1597.1: Standard for Validation of Computational Electromagnetics,Computer Modeling and Simulation," in *59th International Wire and Cable Symposium,* Rhode Island, 2010.

[76] G. Marsaglia., W.Tsang and J. Wang, "Evaluating Kolmogorov's Distribution," *Journal of Statistical Software,* vol. 8, no. 18, pp. 1-4, 2003.

[77] G.Zhang, H.Sasse, L.Wang and A.Duffy, "A statistical Assessment of the Performance of FSV," *ACES Journal,* vol. 28, no. 12, pp. 1179-1186, 2013.

[78] H. Hassani and E. Silva, "A Kolmogorov-Smirnov Based Test for Comparing the Predictive Accuracy of Two Sets of Forecasts," *Journal of Econometrics,* vol. 3, no. 3, pp. 590-609, 2015.

[79] N.Gover, "Two-sample Kolmogorov-Smirnov test for truncated data," *Computer Programs in Biomedicine,* vol. 7, no. 4, pp. 247-250, 1977.

[80] D. Lekomtcev and R. Marsalek, "Evaluation of Kolmogorov - Smirnov Test for Cooperative Spectrum Sensing in Real Channel Conditions," in *22nd Telecommunications Forum (TELFOR)*, Belgrade, 2014.

[81] R. Simard and P. L'Ecuyer, "Computing the Two-Sided Kolmogorov-Smirnov Distribution," *Journal of Statistical Software*, vol. 39, no. 11, pp. 1-18, 2011.

[82] J. Gibbons and S. Chakraborti, Nonparametric Statistical Inference, Marcel Dekker Inc., 2003.

[83] F. Farzad, "Bulletin of Environment, Pharmacology and Life Sciences," *Predicting Global Solar Radiation Using Genetic Algorithm*, vol. 2, no. 2, pp. 54-63, 2013.

[84] W. Vermillion, *End-to-End DSL Architectures*, Cisco Press, 2003.

[85] G. Ballou, Handbook for Sound Engineers, Fourth Edition, Elsevier Inc., 2008.

[86] J.Hayes and P.Rosenberg, Data,Voice and Video Cabling, 3rd Edition, Delmar Cengage Learning, 2009.

[87] P.Golden, H.Dedieu and K.Jacobsen, Fundamentals of DSL Technology, Auerbach Publications, 2015.

[88] J. Cioffi and J.Fang, "A Temporary Model for EFM/MIMO Cable Characterization," in *IEEE 802.3 Standards Contribution*, Los Angeles, 2001.

[89] T. Starr, M.Sorbara, J.Cioffi and P. Silverman, DSL Advances, Prentice Hall, 2003.

[90] W.Chen, Home Networking Basis: Transmission Environments and Wired/Wireless Protocols, 2004, Prentice Hall.

[91] M.Kozak, L.Cepa and J.Vodrazka, "New Values of Cross-Talk Parameters for Twisted Pair Model," *Advances in Electrical and Electronic Engineering*, vol. 8, no. 5, pp. 130-133, 2010.

[92] FLUKE networks, "DSX-5000 CableAnalyzer," FLUKE networks, [Online]. Available: http://www.flukenetworks.com/datacom-cabling/Versiv/DSX-5000-Cableanalyzer. [Accessed 30 August 2016].

[93] FLUKE networks, "Datasheet:DSX-5000 CableAnalyzer," FLUKE networks, [Online]. Available: http://www.flukenetworks.com/doc_links_pdf/en/content/datasheet-dsx-5000-cableanalyzer. [Accessed 3 August 2016].

[94] FLUKE networks, "Versiv Cabling Certification Product Family User's Manual,Versiv Software Version 4.5," FLUKE networks, April 2014. [Online]. Available: http://download.flukenetworks.com/Download/Asset/9828877-e-en.pdf. [Accessed 3 August 2016].

[95] A.Oliviero and B.Woodward, Cabling: The Complete Guide to Copper and Fibre-Optic Networking,Fourth Edition, Wiley Publishing Inc., 2009.

[96] J.Hayes and P.Rosenberg, Data,Voice, and Video Cabling,3rd Edition,, Delmar Cengage Learning, 2009.

[97] ANSI/TIA/EIA-568-B.2-1-2002, *Commercial Building Telecommunications Cabling Standard, Part 2: Balanced Twisted Pair Cabling Components - Addendum 1: Transmission Performance Specifications for 4-pair 100-ohm Category 6 Cabling*, ANSI/TIA/EIA, 2002.

[98] Y.Wang, Fundamental Elements of Applied Superconductivity in Electrical Engineering, John Wiley and Sons, 2013.

[99] A.Semenov, S. Strizhakov and I. Suncheley, Structured Cable Systems, Springer, 2002.

[100] Microsoft, "Add,change,or remove a trendline in a chart," [Online]. Available: https://support.office.com/en-gb/article/Add-change-or-remove-a-trendline-in-a-chart-fa59f86c-5852-4b68-a6d4-901a745842ad. [Accessed 30 August 2016].

[101] K.Stroud and D. Booth, Engineering Mathematics,Fifth Edition, Palgrave, 2001.

[102] S.Guruprasad, A Textbook of Engineering Mathematics -III, New Age International, 2013.

[103] H. G. Sasse, M. M. Al-Asadi, D. Coleby, A. P. Duffy, K. Hodge and A. J. Willis, "A Genetic Algorithm Toolkit for Cable Design," in *50th International Wire and Cable Symposium*, Florida, 2001.